ELEMENTARY DESCRIPTIVE STATISTICS

For Those Who Think They Can't

ARTHUR COLADARCI

THEODORE COLADARCI

Stanford University

WADSWORTH PUBLISHING COMPANY

Belmont, California

A Division of Wadsworth, Inc.

Sponsoring Editor: Roger S. Peterson

Editorial/production services by
Phoenix Publishing Services, San Francisco

Printed in the United States of America

  2 3 4 5 6 7 8 9 19—84 83 82 81 80

Library of Congress Cataloging in Publication Data

Coladarci, Arthur P
    Elementary descriptive statistics.

    Includes index.
    1. Statistics. I. Coladarci, Theodore, joint
author. II. Title.
QA276.12.C64   519.5   79–16643
ISBN 0–534–00782–1

CONTENTS

PREFACE

Textbooks that treat the field of statistics, like the gentle rains from heaven, fall frequently upon the earth. Why another one and why this one particularly? Our somewhat immodest response is that we believe this book fills a need that is not addressed adequately elsewhere. We are certain about our purposes, which can be expressed in terms of the particular audience for whom we conceived and prepared this publication.

Our view is focused on those professional and preprofessional persons whose need for statistical sophistication can be satisfied at elementary levels. This audience is occupationally heterogeneous. It includes persons who are institutional administrators, teachers, lawyers, managers, executives, business persons, civil servants, and nurses, among many others. Many in such occupations, of course, wish or are expected to have statistical competence beyond any definition of "elementary." We do not write primarily for them. (If you are expecting this book to satisfy entirely the prerequisite for an advanced course in statistics, stop here—you have the wrong book.) Our self-imposed restriction to elementary statistics is more severe than others. Except for the final chapter, we deal only with concepts, cautions, and techniques used in *describing* quantitative information and interpreting such descriptions. This restriction is explained in the opening chapter. We hope that you are tempted to read that far.

Our focus on statistical description, we believe, is appropriate and helpful to those who, in the intelligent practice of their roles, are unavoidably confronted with data presentations and with the need to perform elementary descriptive analyses. We also have in mind students in introductory measurement courses, where understanding of descriptive statistics is assumed and where this book may serve as an adjunct to the major textbook.

Among the students and "workers in the vineyards" who fit the foregoing description are many whose earlier training in quantification was weak or is now

v

only a dim memory. If you belong to this populous troubled tribe, please know that this book was planned and developed especially with you in mind. We assume only that the reader is reasonably intelligent, has at least a mild interest, and possesses average patience. Beyond those attributes, some elementary knowledge about numbers will suffice—if you are doubtful on this score, study Appendix I, where we have reviewed some of this prerequisite knowledge. We have simplified as much as possible without courting the dangers of oversimplification. We have introduced deliberate redundancy, risking the impatience of those who "get the hang of it" quickly.

In sum (to use a quantitative term!), this book presumes to be instructive on some elementary statistics for persons in a variety of fields who feel uneasy about their "prestatistical" preparation. We trust that our reach has not exceeded our grasp by too great a margin.

The completion of this modest book proved to be more difficult than we had anticipated. A "simple" treatment of complexity is a far from simple task. If we have succeeded, the accomplishment is shared by many and our thanks go genuinely and gratefully to them all. In the early stages the professional reviewers were: Enoch Sawin, California State University, San Francisco; Peter Apostolakos, California State University, Chico; Donald Rossi, De Anza College; Joseph K. Bryant, Monterey Peninsula College; William Rabinowitz, Pennsylvania State University. We are especially grateful to those who helped us polish the manuscript in its final stages: Walter Riddick, Department of Social Work, Howard University; L. James Tromater, Department of Psychology, University of Richmond; David L. Henderson, Department of Secondary Education, Sam Houston State University; Nissim Shimoni, Department of Educational Psychology, University of North Carolina at Greensboro; Janet Spector, Department of Educational Psychology, Stanford University.

A special smile of appreciation to Valerie K. Familant, Stanford University, for wise criticisms and suggestions on language and form as well as for her perceptiveness in noting implications not intended. To Joanne Shizuru and Gladys Johnson, we are obligated for their competent and patient typing of the several manuscript drafts. Finally, although she may not know it, Jane Coladarci provided significant support, doing the infinity of things she does so well.

Arthur Coladarci
Theodore Coladarci

1

PRELIMINARY ORIENTATION
AND UNDERSTANDINGS

AN ORIENTATION

Our interesting neighbor, Eman A. Ton, recently visited the quaint little country Ecalpon. While in Ecalpon, Eman learned many things about Ecalponians. "I can tell you," he told us, "that they are very discourteous people."[1]

Eman's eyewitness testimony involved two processes: (1) he noted and summarized the courtesy behavior of a particular group of Ecalponians ("I ran into approximately seventy-five," he said—in what we trust is only a metaphor); (2) he generalized his findings to all Ecalponians. These two processes, found in the everyday behavior of all humans, are represented in the two general domains of statistical analyses: description and inference.

DESCRIPTIVE STATISTICS

The data we gather, or are confronted with, arise from a particular and finite group of people or things—for example, the career aspirations of 203 members of the 1978 graduating class at Woodside High School; the daily liquid intake of a particular patient during a given month; the bid-ask performance of Goodall common stock over the last two-year period. **Descriptive statistical methods** are those procedures and rationales that are useful in organizing, summarizing, and interpreting the data actually obtained from the group actually observed. This book deals with that branch of statistics.

---

[1] We mean no disrespect to Ecalponians; indeed, some of our best friends are Ecalponians. Please know that Eman, who does not speak Ecalponi, defines discourteous as "ignoring me." He is a very big man in his very small hometown.

STATISTICAL INFERENCE

Suppose that Eman A. Ton had said, "I met approximately seventy-five Ecalpo-nians, most of whom were discourteous." (You will now recognize this as a descrip-tive statement.) We might have merely gently prodded him on the implications of his definition and the validity of his observations. Good neighbor Eman, however, did not stop with the observed data; he passed profoundly beyond—"Ecalponians are discourteous."

Eman inferred something about all Ecalponians from his knowledge about a particular group of Ecalponians, assuming that those he observed were reasonably similar to all Ecalponians in courtesy-behavior. This is to say that our neighbor engaged in **statistical inference**—enthusiastically, if not wisely. The concepts, logics, and techniques of statistical inference (how to infer about populations from samples) are beyond the purposes of this book, except for the general introduction offered in the final chapter.

The distinction between description and inference is logical rather than psychological. It is distinctively human to draw inferences from immediate experi-ences. The responsible person, like all persons, draws inferences all the time. He is distinguishable from others, however, in his appreciation for the frequently fragile connection between fact and generalization and the corresponding caution with which he generalizes beyond what is immediately known. Therefore, although this book deals only with the descriptive aspects of statistics, we cannot stop you from making inferences as you read. Rather, our injunction is always to question yourself on the limits to which a particular set of data probably can apply—a caution that may be guided adequately by the common sense of alert and reasonably intelligent people. Common sense, however, is not a common attribute. When you have fin-ished this book, we invite you to learn some of the formal concepts and techniques used in the sometimes complex but always fascinating domain of inference and probability.

Let us return, for a moment, to Eman A. Ton, who has been waiting patiently with a contrite apology. Having looked over our shoulder as we wrote, he now is persuaded that his observation and inference on Ecalponian discourtesy may have been careless. "Indeed," he says, "for all I know, the people of Ecalpon may be quite courteous as a general rule. All I really can say is that they have some strange habits. For example, Ecalponians tend to walk in single file—at least" (he added cautiously) "the one I saw did."

OVERVIEW

We deal with the techniques of statistical description, bearing in mind the needs and interests of those who frequently must engage data at elementary levels. Some find it necessary and useful periodically to collect, analyze, and present data—that is, to start from scratch. Such persons (for example, educators, administrators, nurses, managers) also join with many more in the need to respond intelligently to the constant stream of descriptive statistics appearing in the general and specialized press and in relevant reports.

We include techniques and procedures most generally used in descriptive analysis and reporting. In Chapter 2, we consider ways of organizing initial raw data in forms capable of analysis and effective graphical presentation. One of the most common descriptors of sets of data, the average, is considered in Chapter 3. Chapter 4 describes methods for determining and reporting the amount of variation represented in sets of data. Many occupations (for example, education, applied social science fields, personnel work, supervision) use methods of expressing relative performance of individuals; Chapter 5 covers those techniques. In Chapter 6 we present some common elementary ways of describing the relationship between two sets of data. Finally, as we noted earlier, we offer a chapter (7) introducing the world of statistical inference, which we hope will be your next level of systematic study.

In this book, we make no reference to the use of computers in computation and analysis. The omission is neither an oversight nor a rejection of the twentieth century. To the contrary, we strongly urge the use of computers for large computational tasks and complex analyses, if computer services and funds are available.

This book, however, is addressed specifically to those whose need to produce descriptive statistics typically involves small sets of data and relatively simple operations. For such small producers, reference to computer services usually is unrealistically expensive; do it yourself is more prudent advice. The chore and tedium of computation are reduced to relative insignificance by using one of the inexpensive electronic calculators, an aid that also reduces computation errors.

The study of statistics requires the learning of some new language—and the use of some old language in new ways. We define these new terms and new usages as we go along. Before entering the next chapter, however, you will find it helpful to understand what statisticians (and scientists generally) mean by the term *variable* and some distinctions among variables.

VARIABLES

Statisticians gather data about some characteristic of a group or phenomenon. For example, data may concern the age, marital status, academic achievement, or political party affiliation of people in a group. Or they may concern the price of automobiles in some grouping of cars. Individuals or objects *vary* with reference to that characteristic (for example, people vary in age, cars vary in price). Thus, the characteristic of interest is called a **variable.** There are different kinds of variables, however, and differing statistical treatments for the different kinds. Variables fall into two broad types: quantitative and qualitative.

Quantitative Variables

If the observed characteristics vary in magnitude, the variable is said to be a **quantitative variable.** Age, to give an easy example, is a quantitative variable because the differences in age are differences in magnitude of that characteristic. Similarly, tolerance for ambiguity is quantitative, in the statistical sense. A quantitative variable, then, is a characteristic that can be ordered into a hierarchy of amount, size, goodness, preference, degree, and so forth.

Quantitative variables fall into two logical categories.

Discrete Variables.   We have a **discrete variable** when the observations are countable units, expressed only in indivisible wholes. Examples of discrete variables are number of children in a family, number of traffic accidents during a holiday, and number of shares traded in a stock exchange in a given period of time. In the case of such variables we cannot imagine any observed values that would not be whole numbers. The number of children in a family, for instance, is 0 or 1, or 2, 3, or 4, or some larger number, and no observation can fall between those values.

Continuous Variables.   A variable is said to be a **continuous variable** when its measurements, at least theoretically, differ by infinitely small values. Consider, for example, the variable *time in running the mile*. The number of possible time values is infinite, is it not? If we are told that the winning time is 4.56 minutes, we understand that the value was expressed only to two decimal places for convenience and because of the limitations of the timing instrument. That is, 4.56 minutes is our *approximate* value and it could never be expressed as an *exact* value, no matter how many decimal places your patience and the accuracy of your clock would permit you to add. Perhaps you now sense why the term *continuous* is used for such a variable—the values fall along a continuing line; to go from one observed value to another, we pass through an infinitely large number of fractional parts of whatever unit of measurement we happen to be using.

Whether a variable is discrete or continuous depends on the nature of the variable rather than on the nature of its measurement. *Arithmetic competence* is called a continuous variable because we imagine such competence as changing continuously rather than by unit jumps. However, when the variable is measured by the number of items correct on an arithmetic test, the measurement is discrete—that is, it changes in units that cannot be smaller than 1. Most of the variables that interest us probably are continuous rather than discrete; they are treated as continuous even if the measurements are discrete, as is frequently the case when the measurement is a counting measure.

Qualitative Variables

Variables that cannot be expressed in quantitative terms are called **qualitative variables.** The observations cannot be said to vary in magnitude; a natural hierarchical ordering is not possible. Consider the instance of *marital status,* legally defined. The observations on this variable take the form of single, married, or divorced. These categories do not differ in amount or magnitude; they differ in kind.

Observations on a qualitative variable are identified as independent categories or names rather than as quantities. (These observations are sometimes referred to as *categorical* or *nominal*.) *Political party affiliation* is a qualitative variable; observations on this variable are categorized or named as Democratic, Republican, and so on. It is a frequent practice to use numbers to name observations on a qualitative variable. For example, the sexes of individuals might be recorded by using the numbers 1 for female and 2 for male. Such numbers, however, are to be recognized simply as names rather than as quantities. You would not, for instance, think of the numbers on the backs of football players as values on a quantitative scale.

STATISTICS AND SANITY

The real world is made up of people behaving and things happening. Statistics and statistical techniques are *not* the reality. They are simplifiers of reality. Statistics reduces the noisy, complex confusion of specific events to more easily grasped descriptions and orderings of those events. The statistician (like the poet and artist) tries to interpret reality in comprehensible forms, summaries, and analogies.

Statistics and other simplifiers are valuable and necessary to intelligent behavior. However, in learning about and using statistics, you should resist the temptation to believe that the reality is as simple as the simplifier.

In simplifying and ordering reality, statistics also frequently distorts it—as a map or even an aerial photograph, no matter how carefully prepared, cannot exactly match the terrain. It is also, and unfortunately, true that the distortion may be deliberate in some cases. "Figures don't lie," said the puckish sage, "but liars sometimes figure!"

An understanding of statistical techniques is the best protection against abusing or being abused by them. The chapters that follow attempt to introduce you to statistics. With some attention and patience, you may discover that the subject is sensible, useful, and perhaps even fun.

ORGANIZING AND PRESENTING DISTRIBUTIONS

Data, when they[1] first appear, comprise a set of individual observations (e.g., ages, scores, incidents). Unless there is only a very small number of cases, we must organize and present them in a way that makes them easily read. In the case of *quantitative* data, some organizing of the individual bits of data is also required for the convenience of further statistical analyses, when analyses are done by hand rather than by use of computer. Organization of original data also is required for presentation in summary form.

QUALITATIVE DATA

We deal with qualitative data first because the treatments are simpler than those for quantitative data.

In order to gather data for illustration, we make you a member of the Busciardo Club,[2] a friendly group of eighty people. For reasons we cannot imagine, you wonder about the religious preferences of club members (a variable you should recognize as qualitative). Patiently, you ask each member to indicate his or her religious preference in terms of categories that happen to interest you: Catholic, Protestant, other denomination, no preference. The immediate result of our survey is a collection of eighty individual responses, which must be organized in some way. In the case of a qualitative variable, the basis for organization is simply determined—we use the qualitative categories themselves, showing the number of observations that fall into each category. When this is done, we find that the Bus-

---

[1] If you are a Latin purist (and we are old fashioned in this matter), you will use the term *data* as a plural noun; the singular is *datum*.

[2] We trust that no such club exists outside our imagination. The word is Italian; to honor its ethnicity, you should pronounce it "Boo-SHIAR-doh"; to discover its meaning, consult an appropriately ethnic friend.

ciardo Club consists of 16 Catholics, 20 Protestants, 16 Other, 20 No Preference, and 8 who said, "none of your business"—or a less restrained equivalent.

In order to provide an easier reading of qualitative data, results may be presented by means of a table or some type of a chart.

Tabular Presentation

A table is formed by listing the organizing categories in the first column and the numerical data in one or more adjacent columns. Table 2.1[3] presents our fictitious survey results in this form.

The table introduces two symbols commonly used in descriptive statistics and employed throughout this book. The **total number of observations** is symbolized by the upper-case letter N; the number of cases associated with each subcategory is called a **frequency** and symbolized by f. We have translated each frequency into a percent, an easily communicable way for indicating relative frequency. The categories in the left-hand column may be listed in any order; there is no correct order on a qualitative variable.[4] Because Table 2.1 is a way of showing the distribution of the various frequencies across the categories of the variable, it may be called a **frequency distribution,** a term you will see often in descriptive statistics.

Graphic Presentation of Qualitative Data

A commonly employed mode for graphic presentation of qualitative data is the *bar chart* or *bar graph.* As the illustration in Figure 2.1 suggests, construction is a

TABLE 2.1   Religious Preference of Busciardo Club Members ($N = 80$)

Religious preference	f	%
Catholic	16	20
Protestant	20	25
Other	16	20
No preference	20	25
"None of your business"	8	10
Total	80	100

[3] Practices used in titling tables very somewhat. In this book, we follow the convention of placing the title above the table, aligned at the left with the table and with the words capitalized. For graphic presentations (called *figures*), the title is placed below and written as a sentence. You are free to depart from our convention, of course, as long as you are clear and consistent.

[4] However, some prefer to list variables alphabetically to avoid any implication of bias.

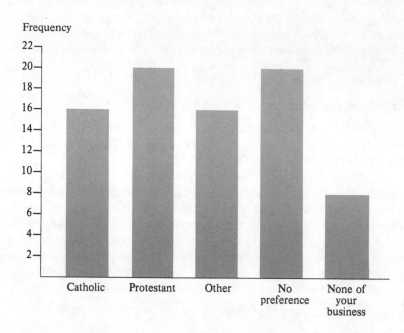

FIGURE 2.1   Bar chart showing the religious preference of Busciardo Club members.

simple matter. The possible frequencies are marked off on the vertical axis,[5] and the categories of the variable are identified on the horizontal axis. A rectangle is erected over each category with the height of the rectangle extending to the appropriate frequency for that category.

Figure 2.1 is easily read by most people. If it is important that the reader be able to determine the *exact* frequency represented by each rectangle, you should construct and present the bar chart on graph paper with sufficient graduation of lines to give that information easily and visually.

Bar charts can be shown with the axes reversed, placing the frequencies on the horizontal axis. Let aesthetics and economical use of space be your guides in this decision. You often have seen bar charts which substitute drawings for the rectangles, using something that pictures the particular variable. In a chart showing automobile sales, for example, one might use outlines of a car—say, one car for every 100,000 sold. A human figure may be used in a population chart, with the height of the figure corresponding to the frequency.

The *pie chart,* named for obvious reasons, is a popular graphic expression of qualitative data, particularly in the news media and public reports. As you can see

---

[5] The usual convention is to mark off the vertical axis with the values increasing from bottom to top.

in Figure 2.2, each of our categories is expressed as a slice of the pie. To construct a pie chart accurately, you need a protractor and the knowledge that a circle represents 360 degrees. We apportion the degrees in the circle according to the percentage represented by each category of the variable. For example, the slice for "none of your business" (Table 2.1) covers 36 degrees of the circle (10 percent of 360).

The pie chart provides an easily interpreted visual comparison of categories. When there are many small categories, however, both construction and interpretation become unrealistic—the slices become too small. Under such circumstances, it is more appropriate to use the bar chart for graphical presentation.

QUANTITATIVE DATA

A quantitative variable, you will recall, is one in which the observed variations are differences in magnitude. Here the organizational task is a little more difficult than for qualitative variables. In the latter (as we have just seen) the subcategories of the variable are given to us (for example, types of religious preferences, or married–nonmarried, or male–female). When we treat a quantitative variable, however, we must create these subcategories.

Organizing Quantitative Data into Intervals[6]

Let us suppose, for simple illustration, that we are interested in the size of classes (number of pupils per class) in the Swampwater Elementary Schools. We find that

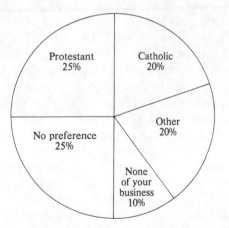

FIGURE 2.2   Pie chart showing the religious preference of Busciardo Club members.

[6] In this text, we refer to subcategories of quantitative data as *intervals*. If you are reading elsewhere, you will encounter other labels, for example, class size, score limits.

there are twenty classrooms in the system ($N = 20$), the sizes of which are as follows (listed in the order in which they were obtained):

$$28, 29, 28, 27, 25, 33, 29, 21, 36, 32,$$

$$33, 32, 29, 38, 35, 25, 31, 26, 27, 28$$

The quantitative variable (class size) ranges in magnitude from 21 through 38. How can we present these data? We can, of course, make a column of all possible class sizes from 21 through 38 and indicate the frequency with which each occurs. Suppose, however, that we are not interested in that much detail, that we will be satisfied with some lumping of class sizes into fewer intervals of class sizes. How many intervals? How wide should each interval be? There is no correct answer to these related questions. Several arrangements are possible as long as all intervals are the same width and each observation can be placed into one, and only one, of the intervals.[7] Because the *midpoint* of each interval data will be used to stand for all observations in that interval, it is convenient to select an odd rather than even interval width (thereby avoiding a midpoint that is a fractional part of the unit observation).

In the case of the Swampwater data, we might decide to divide the range of possible class sizes into interval widths of 5. Table 2.2 is the result of such a decision. The lowest interval, 20–24, covers class sizes 20, 21, 22, 23 and 24— i.e., its width is 5. This interval was selected as the lower one because it will account for the smallest classroom size, which is 21. However, several other five-wide intervals would have met this requirement equally well: 17–21, 18–22, 19–23, or 21–25. Any one of these would do.[8]

Real Limits for Continuous Data. We selected, for our initial illustration, data that represent a discrete series (number of pupils per classroom). The measurements are exact (if we counted correctly). A particular class size is exactly 25 and the next possible size is exactly 26—no in-between values are possible. Therefore, we are not puzzled by the fact that in Table 2.2 the lowest interval ends with 24 and the next interval begins with 25. No observation was skipped in going from 24 to 25. Observations on a continuous series, on the other hand, can be only approximate no matter how precise the measurement may be. If someone tells you that his height is 69 inches, you probably assume that he is reporting to the nearest inch. This means that his exact height is at least 68½ inches but not as much as 69½ inches. If that is too imprecise for your taste, you might ask him to describe his height to the nearest ½ inch, to which he may respond with "69½ inches." You now know that this exact height is at least 69¼ inches but is not as much as 69¾ inches. And so on, ad

---

[7] In formal terms, intervals must be *exhaustive* and *mutually exclusive*.

[8] In Chapter 3, you learn about some assumptions that must be made about how the observations in an interval are distributed throughout the interval. That knowledge will further influence your choices among possible starting intervals.

TABLE 2.2 Frequency Distribution of Class Sizes in Swampwater Elementary Schools (N = 20)

Interval	f
35–39	3
30–34	5
25–29	11
20–24	1
Total	20

infinitum (literally). An observation on a continuous variable represents a *range* of theoretically possible, exact observations, with the particular observation being in the middle of the range. The limits of this range we call the **real limits.** Hence, the real limits of a score of 2 (in a continuous series) are 1.5–2.5. The lower limit is halfway between 1 and 2; the upper, halfway between 2 and 3. Similarly, the real limits of a score of 15.4 are 15.35–15.45.

Perhaps you already see that, with continuous data, the grouping intervals themselves also have real limits. Let us clarify and illustrate this with some further information about Swampwater. We have the grade-point averages (GPA) of the 1976 Swampwater High School graduating class (N = 19)—a very small school! Computed to the nearest one-tenth, these GPA's are:

$$1.7, 1.9, 2.0, 2.0, 2.4, 2.5, 2.5, 2.6, 2.7,$$

$$2.7, 2.7, 2.8, 2.8, 2.9, 2.9, 3.0, 3.1, 3.3, 3.4$$

We see that the GPA's range from 1.7 through 3.4, a distance of approximately two GPA units. If we made our intervals 0.5 wide, we would have four intervals, which strikes us as reasonable. Let us choose 1.5–1.9, which will include the lowest GPA in our data.[9] However, 1.5–1.9 indicates the interval in terms of the actual observation units (tenths), which we call *apparent limits*. Expressed in real limits, the interval is 1.45–1.95. This means that the interval represents all possible GPA's that are at least 1.45 and not as large as 1.95. What are the real limits of the next highest interval? It must begin at 1.95 and go to (but not include) 2.45. The completed frequency distribution for Swampwater GPA's is shown in Table 2.3 with intervals shown in real limits.

[9] Why 1.5–1.9 rather than other intervals that also include the lowest GPA of 1.7? Why not, for example, 1.3–1.7, 1.4–1.8, 1.6–2.0, or 1.7–2.1? The question is appropriate but hold it until you read the final chapter section, "More on Ambiguity and Error."

TABLE 2.3 Frequency Distribution of GPA's of Swampwater High School 1976 Graduating Class ($N = 19$)

GPA (real limits)	f
2.95–3.45	4
2.45–2.95	10
1.95–2.45	3
1.45–1.95	2
Total	19

As an exercise, you might construct a frequency distribution with different interval widths and starting points.

Midpoints of Intervals. A practice in descriptive statistics for certain types of computation is to let the value of the middle of an interval (the **midpoint**) represent all the measurements that are grouped into that interval. This midpoint of the interval may be determined by adding half of the interval width to the lower real limit of the interval. For example, the interval width in Table 2.3 is .5, half of which is .25. If we add .25 to the lower real limit of each interval, we have the interval midpoints. Hence, the midpoint of the first interval is $1.45 + .25 = 1.7$. You must remember that half of the interval width is added to the lower *real* limit, not the apparent limit.

Why all the fuss about real limits? If we ignored real limits, the frequency distribution in Table 2.3 would have been the same. The importance of knowing real limits for intervals lies in their use in constructing graphic representations and in computing central tendencies, and percentiles, as you will see shortly. Indeed, it usually is not necessary to show the real limits in tabular frequency distributions. It is important only that you can find them when necessary.

Ambiguity and Error. There are two general advantages in grouping original data into intervals.

1. The resulting frequency distribution more clearly shows the general pattern or shape of the data—tendencies difficult to sense in the collection of original measurements.

2. Calculations of averages, variabilities, and relationships are accomplished more easily from grouped date than from original measurements.

We pay a price for this convenience, however. When we put a measurement into an interval (that can include more than one measurement), we lose the identity of that measurement. Look again at Table 2.3. The bottom interval includes two GPA's. What *were* those GPA's? There is no way of telling from the table, other than that they lie between 1.45 and 1.95. That is, we have introduced some ambiguity by grouping our data. Error is present also, depending upon what assumption you make about where those two GPA's might lie. If, for example, you assumed that the midpoint of that interval (which is 1.70) fairly represents the two observations, you would be wrong in this case. The two GPA's actually were 1.7 and 1.9; they are neither at the midpoint nor do they average to the midpoint.

You therefore can see that error is introduced when a set of data is grouped into intervals. However, all is not lost. We know that the amount of **grouping error** is related to the size of the interval. For any set of data, this error becomes smaller as the width of the interval is decreased. Conversely, grouping error increases as interval width increases. With larger intervals, the grouping is coarser—hence, the term *error due to coarse grouping*. To put this another way, the larger the number of intervals you use in organizing data, the smaller the error due to grouping. Of course, as the number of intervals is increased, a point of diminishing returns is approached: the advantages of grouping disappear if we have as many intervals as there are possible observations. The general practice is to strive for at least ten intervals—and more when N is large and the observations cover a wide range.

You now can see that the Swampwater examples we used earlier are poor in this respect. We used those data merely to provide a very simple example, not to set a bad example.

Graphical Presentation of Quantitative Data

There are several standard methods of presenting quantitative data in graphic form. Most common are the *histogram,* the *frequency polygon,* and the *time graph.* In some fields of work, particularly education, the *cumulative relative frequency polygon* also is found useful.

To explain and illustrate each of these methods, we return to the members of the Busciardo Club, whose religious preferences we already know and whose body weights we now discover. We have measured the weight of each member of this cooperative group in kilograms.[10] We examine the eighty weights, which ranged approximately from 51–93 kilograms, and decide on an interval width of five. Table 2.4 is the result in tabular form.

In this table, we show real interval limits and midpoints because we need them in graphical representation. With Table 2.4 in hand, we can proceed to the methods for presenting this frequency distribution in graphic form.

---

[10] If you are not yet comfortable with the metric age, 1 kilogram is 2.2 pounds, approximately.

TABLE 2.4 Weights (in kilograms) of Busciardo Club Members

Weights	Interval real limits	Midpoints	f
90–94	89.5–94.5	92	10
85–89	84.5–89.5	87	17
80–84	79.5–84.5	82	11
75–79	74.5–79.5	77	6
70–74	69.5–74.5	72	5
65–69	64.5–69.5	67	3
60–64	59.5–64.5	62	7
55–59	54.5–59.5	57	13
50–54	49.5–54.5	52	8

The Histogram.   Figure 2.3[11] is a histogram of weights of Busciardo Club members. The real interval limits are marked along the horizontal axis with equal distances between them. On the vertical axis, we marked off a scale of frequencies and the frequency for each interval is shown by the height of a rectangle drawn over the interval. Construction of histograms is a relatively simple matter (especially so if you use graph paper). The result is an effective visual portrayal of a frequency distribution. (Incidentally, can you guess from Figure 2.3 whether the Busciardo Club is all male, all female, or mixed sex?) If the real limits are marked off carefully, the relationship between frequency and area is perfect. The proportion of the total histogram area represented in a rectangle is equal to the proportion of the total number of cases (N) represented by the frequency in that interval.

The construction of any graph involves an annoying decision and an unsolvable problem. How long should the vertical and horizontal axes be, relative to each other? Different relative lengths of the two axes result in different visual impressions of any given frequency distribution. Indeed, an unprincipled graph maker can distort the visual impression in a particular way by manipulating the relative lengths of the two axes. (Later, in discussing time series, we illustrate how easily this may be done.) There is no correct relationship between lengths of vertical and horizontal axes. Some statisticians suggest, as a rule of thumb, that the vertical be approximately two-thirds to three-fourths of the horizontal. We suspect that this rule (or any other) represents an aesthetic preference rather than a technique for controlling liars. Let your conscience be your guide when you construct a graph—and be alert when you examine one.

[11] As was pointed out earlier, the usual practice is to show the scale values increasing upwards on the vertical axis. In the case of the horizontal axis, the values usually are shown to increase from left to right.

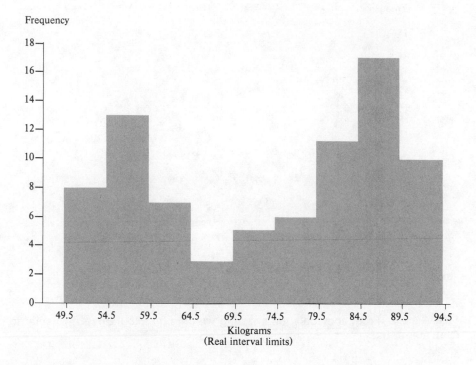

Frequency

FIGURE 2.3   Histogram showing weights (in kilograms) of Busciardo Club members.

The Frequency Polygon.   Using the same axes, we can visualize the frequency distribution by placing a dot over the midpoint of each interval and connecting the dots with straight lines. A graph in this form is called a **frequency polygon.** Figure 2.4 is the frequency polygon for weights of denizens of the Busciardo Club. On the horizontal axis, the midpoints of the intervals are marked off, equally spaced. Above each midpoint, we mark a dot opposite the appropriate frequency given on the vertical axis and we connect the dots with lines. In order to complete the polygon, we must do one more thing. Notice that the lowest and highest midpoints shown in Figure 2.4 represent new class intervals. We did not show them in Table 2.4 or in the histogram because no cases fell into them (the frequency for each is zero). It is conventional in constructing frequency polygons to bring the line to zero at each end (that is, to close the graph)—hence the inclusion of an additional midpoint at each end of the horizontal axis.

Although Figure 2.4 shows midpoints, you will have much company if you mark off in real limits instead, as long as you place your dot above the midpoint of each interval.

Frequency polygons and frequency histograms serve the same purpose and in

Frequency

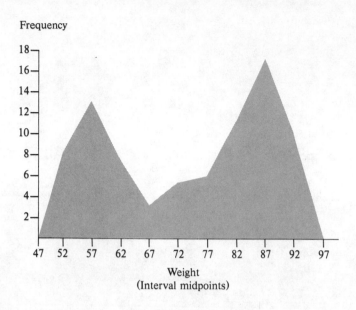

Weight
(Interval midpoints)

FIGURE 2.4   Frequency polygon showing weights (in kilograms) of Busciardo Club members.

only slightly different ways. Indeed, you can convert a histogram into a polygon merely by placing dots at the top of each rectangle, connecting the dots, and bringing each end down to zero frequency. It would be difficult to argue that one is more effective than the other, although some feel that a frequency polygon reveals more clearly the underlying "shape" of a distribution (a concept to be discussed later in this chapter).

Cumulative Relative Frequency Graph. It is sometimes useful or necessary to identify the proportion of a group who exceed or fall below any given value on the quantitative variable. When N is large, a time-saving and easy method for making such estimates is found in **the cumulative relative frequency graph,** a mouthful of a title but a simple procedure.

The salaries of public school teachers in Taudion provide convenient data for illustration. We first organized the salaries of the 220 Taudion teachers into class intervals, indicating real limits because we must use these in the graph. (Salaries were obtained to the nearest dollar.) The first three columns of Table 2.5 show the intervals (apparent and real) and the frequencies—that is, the familiar frequency distribution in tabular form. The remaining columns are new. The fourth column is labeled "Cum. f," which abbreviates *cumulative frequency* and is produced literally by cumulating the interval frequencies from the bottom up. Each interval

TABLE 2.5   Salaries of Teachers in Taudion Public Schools

Apparent interval (dollars)	Real limits (dollars)	f	Cum. f	Cum. rel. f
19,750–20,999	19,749.5–20,999.5	8	220	1.00
18,500–19,749	18,499.5–19,749.5	13	212	.96
17,250–18,499	17,249.5–18,499.5	15	199	.90
16,000–17,249	15,999.5–17,249.5	18	184	.84
14,750–15,999	14,749.5–15,999.5	22	166	.75
13,500–14,749	13,499.5–14,749.5	28	144	.65
12,250–13,499	12,249.5–13,499.5	29	116	.53
11,000–12,249	10,999.5–12,249.5	32	87	.39
9,750–10,999	9,749.5–10,999.5	28	55	.25
8,500– 9,749	8,499.5– 9,749.5	27	27	.12

frequency is added to the preceding frequency until we arrive at the top most interval, where the cumulation is equal to N (in this case 220). Before going on, let us see how we read the "Cum. f" column. The bottom-most Cum. f is 27. This is also the number of cases in that interval, of course. From the earlier discussion of real limits, you should understand that these twenty-seven salaries are between $8,499.50 and $9,749.50. Now what about the next higher interval? The Cum. f column shows 55 for that interval, meaning that of the total N, 55 teachers have salaries below $10,999.50 (the upper real limit of that interval). This should be clear; if not, look at it again slowly.

The last column of Table 2.5 is prepared by converting each cumulative frequency to a proportion of the total N. For example, for the bottom interval, the cumulative relative frequency is $27/220 = .12$. The uppermost relative frequency must be unity, that is, $220/220 = 1.00$. You probably have anticipated how this column is read. The cumulative relative frequency for each interval tells us what proportion of salaries fall below the upper real limit of that interval—that is, fall in that interval and all intervals below it. Hence, we can say that .65 (or 65 percent) of all Taudion salaries are less than $14,749.50.

Now, to construct a graph for cumulative relative frequencies, examine Figure 2.5, which is the cumulative relative frequency graph for Taudion salaries. Real limits of salary intervals are marked off on the base line. The vertical axis gives us a scale of proportions between .00 and 1.00. Over the **upper real limit** of each interval, we placed a dot showing the cumulative relative frequency at that point and we connected those dots with straight lines.

A cumulative relative frequency graph permits us to obtain, visually, approximate answers to questions about relative frequency. Suppose, for example, we wished to know the proportion of Taudion teachers whose salaries are less than $12,875 per year. From our grouped data, there is no way of getting an exact answer to this question. A usable approximation, however, can be read from Figure 2.5,

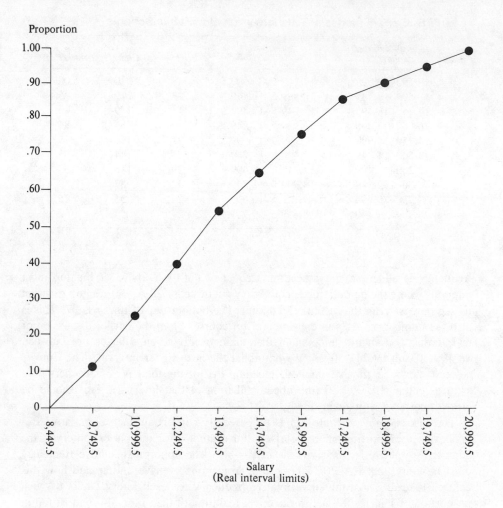

Proportion

FIGURE 2.5   Cumulative relative frequency graph of salaries in Taudion Public Schools.

with a few numerical gymnastics. We read along the horizontal axis and see that $12,875 falls between 12,249.50 and 13,499.50. Easy calculation[12] tells us that $12,875 is almost exactly halfway between these two points. From the halfway point, go perpendicularly up to the frequency curve and at that point go horizontally across to the vertical axis, where you can approximate the cumulative proportion as .46. We can say that approximately 46/100 (46 percent) of Taudion teachers have salaries less than $12,875.

[12] The width of the interval is 1250, the value in which we are interested (12,875) is about 625 greater than 12,249.50. Our salary, therefore, lies about 625/1250 of the distance between 12,249.50 and 13,499.50—or halfway.

Another type of question also can be answered from such a graph. What, for instance, can we say about the salaries of the bottom 25 percent of these teachers? In this case, we begin with the vertical axis (the cumulative proportions) and find .25. That proportion lies between .20 and .30. We determine that .25 is halfway between those two values and we go horizontally from that halfway point to the curve; from the intersection, we go down vertically and note where we strike the horizontal axis, which appears to be at approximately $11,000. An answer to the question, therefore, can be that the bottom 25 percent of Taudion salaries are below $11,000, give or take a little.

Why do we insist on using the qualifier "approximate"? There are two sources of imprecision. One is obvious enough: the probable inaccuracies in visually intersecting the cumulative curve and the axis when we read for our answers. We can improve our approximation, in this respect, by using graph paper and plotting more points on the two axes. The other source of error arises from the necessary assumption that the cases in each interval are distributed uniformly throughout the interval. The assumption is necessary because without it, we would not be able to make estimates for values that did not fall exactly on the plotted real limits. In this respect, approximations are improved as the size of class intervals is reduced. We will close this chapter with more discussion on this matter.

This type of graphical representation, the cumulative relative frequency graph, has useful and frequent application in education, employee selection, and personnel evaluation. We return to it in Chapter 5, where techniques for deriving scores are discussed.

Time Graphs.   The purpose here is to show graphically how observations on a variable change over periods of time. The time graph is worth attention because change over time is of frequent interest. Another reason for presenting it here is that the time graph easily lends itself to distortion for effect. It must be read with great care, despite its simplicity.

To illustrate the construction of a time graph, we use some information given in a recent federal report.[13] Beginning with the 1965–1966 school year, and for alternate years thereafter, the expenditures per pupil in the public schools in the United States were as follows: $537, $658, $816, $990, $1,147, and $1,409. (The last figure, for 1975–1976, was a projection.) To graph these expenditures over the time period, we mark off the time intervals on the horizontal axis and expenditure per pupil on the vertical axis. Then we indicate the expenditure for each year either by a rectangle over that year or a dot, connecting the dots with lines. We use the latter form, quite arbitrarily. However, to illustrate the "distortability" of time graphs, we have constructed two different graphs using the same data exactly. Look at Figure 2.6. In both time graphs, the same per pupil expenditures are plotted, and correctly. The top illustration conveys the impression that pupil costs rise, but not at a fast rate. The bottom graph seems to tell us that the rise in pupil costs is fantastic. The

[13] The National Center for Educational Statistics, in its report, *The Conditions of Education,* 1976 edition, p. 258.

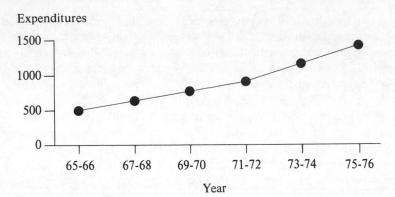

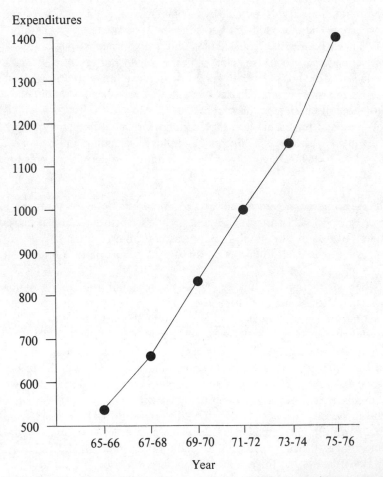

FIGURE 2.6 Two graphic presentations of the same data on annual expenditures per pupil in U.S. public elementary and secondary schools, 1965–1976 (only alternate years).

two different pictures are produced by changing the lengths of the axes, stretching and curtailing to create the desired effect. As we noted earlier in this chapter, there is no antidote for distortion other than vigilance in the consumer and good conscience in the producer.[14] The fact is, of course, that any graph you construct will convey a particular picture.

SHAPES OF DISTRIBUTIONS

One way of describing a set of observations is to describe the general **shape** formed by the frequency histogram or polygon. Knowledge about the shape of the distribution also gives us further information and cues about the variable and the group on which observations have been made. Furthermore, the shape of a distribution has bearing on intelligent interpretation of the data and, in some cases, on what kind of further analyses are appropriate or possible.

Although distribution shapes can be described in many ways, the following typings are the most common.

Symmetrical Distributions

In the upper left portion of Figure 2.7 you see a distribution (a) that shows the left and right halves of the polygon as mirror halves of each other. This is a **symmetrical distribution.** It is rarely the case that a distribution of actual observations is precisely symmetrical, of course. A tendency toward symmetry, however, is found frequently on many variables if N is large. Approximate symmetry, for example, is characteristic of the distribution of the ages of pupils at a given grade level and the distribution of intelligence test scores in the general population—to give two quite different examples.

Skewed Distributions

When a distribution departs markedly from symmetry, it is said to be a **skewed distribution.** These are distributions in which the observations tend to fall more toward one end of the horizontal axis than toward the other. In Figure 2.7 distributions (b) and (c) show skewness. In the case of (b), the higher frequencies are toward the right and the converse is the case in (c). The convention for labeling the direction of skewness is, unfortunately, confusing to the beginner. The direction of the skew is identified by the side of the distribution where the smaller frequencies stretch out (the side farthest away from the peak of the distribution). Following the custom of calling the left end of any scale the negative end, (b) in Figure 2.7 is described as **negatively skewed**—or, if you wish, you may describe it as **skewed to**

---

[14] For an informative and interesting discussion of this general topic, see Darrell Huff, *How to Lie with Statistics* (New York: Norton, 1954) and Stephen K. Campbell, *Flaws and Fallacies in Statistical Thinking* (Englewood Cliffs, New Jersey: Prentice-Hall, Inc., 1974).

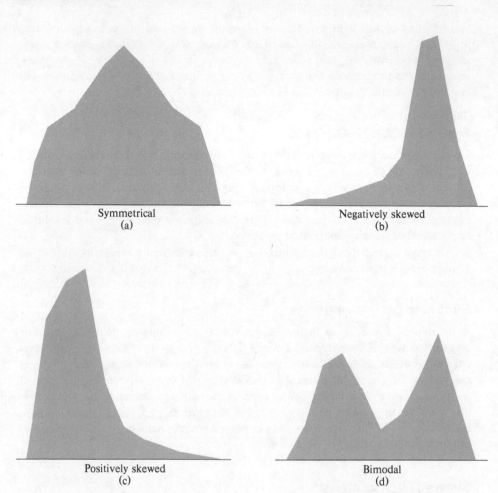

Symmetrical
(a)

Negatively skewed
(b)

Positively skewed
(c)

Bimodal
(d)

FIGURE 2.7   Frequency polygons illustrating different shapes of distributions.

the left. Conversely, the distribution in (c) is said to be **positively skewed** or **skewed to the right.**

Markedly skewed distributions are not uncommon. Indeed, for many variables, you can predict the existence of skewness and direction of skewness. For example, you will not be surprised to learn that the distribution of scores on an easy test is negatively skewed—or that the distribution of weights of jockeys is skewed positively.

Sometimes the skewness is so extreme that we give it a special name, **the J curve,** because the shape of the distribution is something like the letter J. J curves usually result when the behavior being observed is strongly influenced by a social norm. Suppose, for example, that we observed the behavior of motorists at a stop sign, classifying each as: came to a full stop, slowed almost to a stop, reduced speed slightly, no reduction in speed. (You may recognize these as intervals for a quantitative variable.) What would the frequency distribution look like? We bet that it would fit the term *J curve* (although, in this case, a backwards J).

Bimodal Distributions

Frequency distributions usually peak only at one point and fall off gradually toward each end of the horizontal axis. Some data distributions show more than one peak (or, as we shall call it, **mode**). If a distribution has two modes (peaks), it is called a **bimodal distribution.** You have already seen an instance in Figure 2.4 where the weights of Busciardo Club members show two peaks or modes. Distributions also may be **trimodal** (or more), but their occurrence is rare.

When you find bimodality in a distribution you can be fairly sure that the total group of cases comprises two identifiably different subgroups. That is, a bimodal distribution suggests that two overlapping but modally different distributions are present. Earlier, we asked you if you could guess whether the Busciardo Club was unisex or mixed. If you guessed mixed sex because of the two modes, you already fully understand the phenomenon. To confirm it and to illustrate the implications of bimodality, we show you, in Figure 2.8, the weights of Busciardo Club members separately by sex. As you can see, there are indeed two groups (males and females) that differ systematically with respect to weight. Perhaps you have already sensed

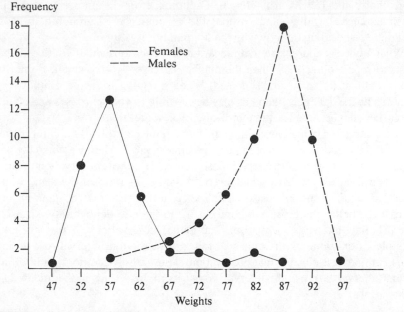

FIGURE 2.8 Frequency polygons of weights of female and male members of Busciardo Club.

that bimodal distribution also may be symmetrical or skewed. On close inspection, for example, the bimodal distribution in Figure 2.4 can be seen to be somewhat skewed to the left.

MORE ON AMBIGUITY AND ERROR

If you have come thus far on our little journey into the domain of descriptive statistics, you have seen that certainty is not on any signpost. Those unhappy souls who cannot tolerate ambiguity and error will find no refuge in statistical analyses—nor anywhere else in the rational world.[15] Any observation on a quantitative variable, no matter how finely calibrated our measuring instrument may be, is an approximation. In specifying our unit of measurement (nearest inch, nearest tenth, and so on), we are indicating the degree of error we are willing to tolerate for the purposes at hand. When we group our observation into intervals, we introduce further ambiguity and possible error—a sacrifice we make in the interests of convenience in computation, presentation, and general description of distribution characteristics.

We have said that the degree of ambiguity and possible error is related to the width of class interval we select. The larger the interval width, the coarser the grouping. As interval widths become smaller for any given set of data, *grouping error* is reduced. Consider, for example, the smallest possible interval width, which would be the unit of measurement being used. In such a case, each possible score becomes its own interval, that is, the midpoint of the interval is the only possible score that could fall in that interval. Grouping error disappears in that smallest possible interval width—and so does the convenience of grouping, because we would then be dealing directly with the original observations.

What advice is appropriate with regard to the decisions about width of intervals? First, to repeat ourselves, when the number of cases is very small, do not bother with grouping; no advantage is gained. When grouping is useful, employ as many intervals as the data permit and as may be sensible to your purposes and audiences. We earlier suggested at least ten intervals. Even more, if you can.

Grouping error in the lowest interval can be reduced explicitly if careful attention is given to the question of where to begin that interval. Earlier, we noted the obvious requirement that the lowest interval must include the lowest observation in our set of data. However, as you have seen, several possible intervals of the same width can meet that requirement. For reasons that easily can be appreciated, grouping error in at least the bottom interval is minimized if the interval deliberately is selected so that its midpoint tends to represent the actual observations in that interval. For example, if only one observation will fall into the bottom interval, use an interval whose midpoint is close to that observation; if several observations will fall into the bottom interval, arrange the interval so that its midpoint lies in the middle of those observations.

---

[15] If you are one of these unfortunates, you will find companionship with the gentleman who was heard to say, "I thought I made a mistake once—but I was wrong."

No matter what one does with grouped data, however, grouping error is likely to be present. Such error can be avoided only by foregoing the advantages of grouping; that is, by working directly with the original observations. You should not assume that working directly with original observations provides error-free results and descriptions. Two sources of error still remain, even if error due to grouping is avoided. One source, you already know. In the case of continuous quantitative variables, the observations necessarily are approximations only—the observation 4.3 is not, and cannot be, an exact measurement.

The other source of error is found in any type of observation, whether continuous or discrete: error that arises because the measurement techniques employed are in many ways error producing. That is, beyond the fact that continuous data theoretically can never be exact, our techniques of measurement (including the ways in which persons apply them) have their own imperfections. From what we know about the measurement of intelligence, for example, we would think it silly scientism to compute a Stanford-Binet IQ as 111.51. More familiarly, but for the same reason, you would not be fooled into an assumption of exactness by the statement: "I paced off the distance and found it to be 36.45 feet."

For most purposes, the error introduced by grouping observations is quite tolerable. Do not be afraid of it.

EXERCISES

1. Identify each of the following variables as qualitative, quantitative and discrete, or quantitative and continuous:

   a. Distance between home and work   **f.** Daily library circulation

   b. Occupation   **g.** Age

   c. Sex   **h.** Type of dwelling

   d. Intelligence   **i.** Source of income

   e. Temperature

2. Construct pie charts or bar graphs for the following sets of data:

   a. The Governor of a western state, in his budget proposal for the 1976–1977 fiscal year, indicated that the percentages of revenue from various sources were expected to be

Personal income tax	29.1%
Horseracing fees	1.9%
Highway users' tax	9.5%
Bank and corporation tax	16.4%
Motor vehicle license fees	4.4%
Sales tax	25.0%

(Problem continues on next page.)

Inheritance and gift tax	3.3%
Other sources	10.4%

b. During a recent year in a major U.S. city, 5,000 persons were naturalized (that figure and others are rounded for convenience). Of the total, the number naturalized under each of the legal provisions was as follows:

Under general provision	4,000
Through marriage to a citizen	500
Through citizenship of a parent	250
Under other provisions	250

c. In 1973, according to the U.S. Bureau of Census, states and local governments spent their revenues in this way:

Education	34%
Highways	9%
Health and Welfare	18%
Utilities and liquor stores	5%
Insurance trusts	6%
Other expenditures	28%

d. (Bar graph only.) Shown below are the average pupil-teacher ratio (number of pupils per teacher) in the primary schools of several countries during 1972:

Canada 24.7	France 23.0
Germany (F.R.) 31.0	Italy 20.2
Japan 24.5	Netherlands 29.1
Norway 19.6	United Kingdom 25.6
United States 25.7	

3. If you can accomplish this exercise without visible pain, you probably understand the meaning of real limits. What are the real limits for each of the following measures?

a. 5 inches

d. Age twelve, to the nearest birthday

b. 5.0 inches

e. Eighty, rounded to the nearest ten

c. Age twelve, as of last birthday

4. The Consumer Price Index reflects the cost of consumer goods. The following are the indices in the month of July during the years of the Nixon administration:

Year	Price index (rounded)
1969	110
1970	115
1971	120
1972	125

Construct two different times graphs from these data, one pro-Nixon and one anti-Nixon.

5. As grist for your statistical mill, we provide some fascinating observations, published here for the first (and only) time. To gather these data, we went to Fumone (a very small place south of Rome) and carefully determined the number of pounds of pasta consumed by each of forty families in a one-month period. (We use pounds here to comfort those who are troubled by the metric system.) Here are the forty results, listed in order of magnitude:

> 53, 57, 58, 59, 61, 61, 63, 65, 66, 67,
> 67, 68, 69, 69, 72, 74, 74, 74, 75, 75,
> 76, 76, 76, 77, 78, 78, 79, 80, 81, 81,
> 82, 82, 83, 84, 85, 87, 88, 88, 90, 95

Order these observations into a frequency distribution, selecting an interval width and midpoints you consider appropriate. Include real limits, midpoints, and frequencies in your table. To show that you can do it, also include the cumulative frequency and cumulative relative frequency for each interval.

6. The following is a frequency distribution of distances (in kilometers) traveled daily by high school students in a rural area. Construct a frequency polygon, a histogram, and a cumulative relative frequency graph for these data:

Kilometers	f
30–32	2
27–29	5
24–26	8
21–23	13
18–20	14
15–17	9
12–14	7
9–11	6
6–8	4
3–5	1

7. Can you predict the probable shape of the distribution for each of the following? (Symmetrical? Skewed? If skewed, in which direction? Bimodal?)

   a. Scores of fourth graders on an algebra test designed for eighth graders.

   b. Weights of adult American men.

   c. Distances a softball is thrown by all sixth-grade pupils in a given school.

   d. Scholastic ability scores of graduates from a high-standards college.

   e. The classification of library visitors on the following scale: are silent, whisper softly, whisper loudly, talk normally, shout.

   f. Length, in minutes, of movies in your local theatres.

   g. Age of assistant professors in American universities.

8. The qualitative variable is reading preferences. What is wrong with the following sub-classification: adventure, biography, fiction, nonfiction?

"ON THE AVERAGE"

You have seen that a set of data can be described in terms of the shape of the frequency distribution. From Chapter 2, you have some glimmer that the shape of a distribution is important knowledge—a glimmer that should become a strong ray of light by the end of this chapter. Two additional descriptors minimally are required for intelligent consideration or communication of a collection of data: (1) the central tendency ("average") of the observations and (2) the amount of variation around the central tendency. The first of these is the subject of this chapter; the second is addressed in Chapter 4. It is unfortunate that, although averages are frequently used in popular communication, shape and variation are generally ignored, thereby inviting superficiality and wrong conclusions.

The average of a distribution of observation may be described in several ways, each way implying quite different information—so different, in fact, that we urge you to avoid using the term *average*. By doing this you can avoid ambiguity; you will not misinform, unintentionally deceive, or be deceived.

The three most frequently used measures of central tendency are the *mode,* the *arithmetic mean,* and the *median*. There are other concepts and measures of central tendency, but their applications are too specialized for this general treatment.

THE MODE

Among measures of central tendency, the mode is undoubtedly the simplest. The **mode** of a collection of observations is defined as the one observation that occurs most frequently. The concept was introduced in the previous chapter in the discussion of bimodal distributions. Consider the following measurements,[1] listed in order of magnitude:

$$14, 18, 18, 24, 24, 24, 29, 31, 88$$

---

[1] We produced these data by journeying to the Trinity Alps and asking each of ten backpackers the following question: "How many grizzly bears have you fought off single-handedly in the last seventy-two hours?" These are the results. Whether they reflect the fantasies of the hikers or the aggressiveness of the grizzlies in the area, we cannot and dare not guess.

The mode of this set of measures is 24, which we find to be the measure with the highest frequency. When a set of observations contains more that one mode, each is identified. This measure of central tendency is important in many areas of human activity. In the garment industry, for instance, production is influenced intelligently by knowledge of modal body measurements. Similarly, furniture manufacturers need to anticipate the most frequently occurring body size in designing and distributing their products. And, in the advertising industry, identification of modes of qualitative data is (annoyingly) a way of life; for example, "Caries Toothpaste is the brand most often used by dentists."

The mode clearly is useful when one observation is markedly more frequent than others. Its utility drops off as the frequencies for different observations approach each other. In the extreme case, where all observations occur with equal or approximately equal frequencies, the mode is meaningless. The latter is a rare event; such a distribution is called *flat* or *rectangular,* for reasons obvious when you visualize the shape of its frequency polygon or histogram.

The identification of the mode (or modes) is a simple matter when you have the original individual observations at hand. But can one determine the modal observation when the original observations have been grouped in a frequency distribution? Obviously one cannot identify the true mode without access to the original observation. The best we can do is to identify the modal characteristic of the grouped data by reporting the *modal interval* or the midpoint of that interval. The latter (the midpoint with highest frequency) is usually referred to as the *crude mode.* Thus, from Table 3.1 we can see that the modal interval is 9.5–12.5 kilometers or that the crude mode (the midpoint) is 11 kilometers. In Table 2.4 (or, if you wish, Figure

TABLE 3.1 Distances (in kilometers) Between Home and Place of Work for Taudion Town Employees (N = 104)

Kilometers	Real limits	Midpoints	f	Cum. f
28–30	27.5–30.5	29	2	104
25–27	24.5–27.5	26	3	102
22–24	21.5–24.5	23	4	99
19–21	18.5–21.5	20	8	95
16–18	15.5–18.5	17	10	87
13–15	12.5–15.5	14	16	77
10–12	9.5–12.5	11	21	61
7– 9	6.5– 9.5	8	18	40
4– 6	3.5– 6.5	5	12	22
1– 3	.5– 3.5	2	10	10

2.4) you see a bimodal distribution, with the crude modes at 57.0 and 87.0 kilograms. (What are the modal intervals in that distribution?)

Making statements about the mode from grouped data, however, is a shaky affair. The fact is that the location of the modal interval and crude mode is different for different interval widths. If we organized the same set of observations in three ways (using different interval widths each time), we would obtain three quite different crude modes and modal intervals. Because of this unfaithfulness, statisticians generally have not developed love affairs with the mode from grouped data. Our advice is not to use this measure of central tendency when the true mode can be identified, making two exceptions in the case of grouped data: (1) when the crude mode represents an exceptionally high frequency compared with other intervals or (2) when a very coarse statement of modal characteristics is acceptable for the purpose at hand.

THE MEDIAN

The **median** of a set of ungrouped observations is the middle score, when the observations are arranged in order of magnitude. It divides the observations into two equal halves. When N is an odd number, the median is found by identifying the middle observation, counting up or down the ordered observations. For illustration, consider the following scores ($N = 5$): 8, 10, 15, 16, 20. The middle score is 15; thus, it is the median of these scores. When N is an even number, the median is the value midway between the two middle observations. For example, consider the following scores ($N = 6$): 25, 27, 29, 31, 34, 34. The two middle scores are 29 and 31; the median is 30, the average of the two middle scores (that is, half of their sum).

The Median from Grouped Data

When observations are grouped into intervals, we cannot find the median by the operation described above. The identity of the original observations has been lost. The location of the median, however, is not as much affected by grouping as is the mode. Finding the median from grouped data is an acceptable and universal procedure.

We begin with a slightly different definition of the median: *the median of grouped data is the point below (and above) which exactly 50 percent of the observations fall*. Look at Table 3.1, which shows how far Taudion Town employees live from their place of work. The median distance is the point (on the scale of kilometers) below which one-half of the 104 individual observations fall. We seek the point that divides the distribution into two equal parts, fifty-two cases on each side ($104/2 = 52$). To do this, we computed the cumulative frequencies and they are given in the right-hand column of Table 3.1. Run your eye up the cumulative frequencies. Where is 52 (half of N)? It must lie somewhere in the interval associated with the cumulative frequency 61. That is, the median lies somewhere between the real limits of that interval (9.5–12.5); it is at least 9.5 (the lower limit) but not as large as 12.5 (the upper limit). But *where* is it? What *point* between 9.5 and 12.5 marks off fifty-two cases?

We approach this question by noting that the next lowest interval shows a cumulative frequency of 40. That is, forty of our required fifty-two cases fall below the top of that interval, which is 9.5—also the beginning of the interval in which the median lies. Therefore, we need an additional twelve cases to make the required fifty-two. The interval in which the median lies shows a frequency of 21, so we must take twelve of these twenty-one cases to make our fifty-two. Now we must make an assumption about those twenty-one cases. We assume that the twenty-one cases in the interval 9.5–12.5 are scattered uniformly throughout the interval. Visualize a straight line with 9.5 at its left end and 12.5 at its right; our assumption is that the twenty-one cases represent twenty-one equally spaced points along this line. That being the case, we can say that twelve cases (the number we require to make fifty-two) represent 12/21 of the scale distance between 9.5 and 12.5. Since the distance is 3 (the size of the interval), we can say that the point we seek is 3(12/21) above the lower limit of the interval (9.5). Putting this all together, gives us the following:

$$\text{median} = 9.5 + 3(12/21)$$

$$= 9.5 + 36/21$$

$$= 9.5 + 1.7$$

$$= 11.2$$

The median distance between home and work is 11.2; half of Taudion employees live farther away and half live closer to work.

A summary computational formula for finding the median of grouped data may be written as follows (where the "interval" means the interval in which you know the median lies and i refers to the width of the class interval):

$$\text{median} = \text{lower real interval limit} + i \left(\frac{N/2 - cum\,f \text{ below interval of median}}{f \text{ in interval of median}} \right)$$

Using this formula, we compute the median for Table 3.1 this way:

$$\text{median} = 9.5 + 3 \left(\frac{104/2 - 40}{21} \right)$$

$$= 9.5 + 3 \left(\frac{52 - 40}{21} \right)$$

$$= 9.5 + 3(12/21)$$

$$= 9.5 + 1.7$$

$$= 11.2 \text{ kilometers}$$

The median is not only an informative measure of central tendency; it is also easily understood by most persons. Furthermore, as you will see later, when both the median and arithmetic mean are known, inferences can be made about the shape of the distribution and we can be more intelligent in interpreting either measure. It is a popular and appropriate measure of central tendency for descriptive purposes. For complex statistical analyses and in statistical inference, the median has two limitations. First, it is not as stable as the mean from sample to sample. Second, it is nonalgebraic—the median is a counting measure, ignoring the magnitudes of the scores. The median cannot be used as a factor in a larger formula or other algebraic expression. We repeat, however, that its descriptive value is very great; we strongly recommend its use in analyzing and communicating quantitative data.

Open-Ended Intervals.   Observations sometimes are grouped into what is called an **open-ended distribution,** a distribution where the lower limit of the bottom interval and/or the upper limit of the top interval are not identified. The following is an oversimplified illustration:

Intervals
125 or more
120–124
115–119
114 or less

This is done for any of several reasons. Most often, the end intervals are left open because the extreme scores are few in number and spread out over a large range. The consequence of including all the actual intervals of equal width would be an awkwardly long listing of intervals, many with zero frequencies. Suppose, however, that one wished to determine the median of scores grouped this way. A median that lies somewhere in the open-ended intervals cannot be calculated. If this fact is not immediately obvious to you, review the procedure for determining the median from grouped data. Recall that, in order to locate the median, we must know the width of the interval in which it lies. When the median lies in an interval other than the two extremes, there is, of course, no problem in computing it.

THE ARITHMETIC MEAN

The third measure of central tendency is an old familiar friend. The **arithmetic mean** of a set of observations is the sum of the observations divided by the number of observations. When most people use or hear the term *average,* they have this meaning in mind. Thus you already understand that the arithmetic mean of the values 7 and 9 is found by computing $(7 + 9)/2$, which gives us 8. There are other kinds of means, but the arithmetic mean is used so frequently that when you encounter a reference to the *mean,* you can be sure that it is the arithmetic mean. If another type of mean is meant, it will be specified. We too hereafter refer to the arithmetic mean simply as the mean.

The formula for the mean can be symbolized in many ways. Our preference (a popular one) involves the use of some new symbols. First, we need a symbol to represent the individual observations on a variable. Here we use the capital letter X. Thus, for example, instead of identifying our data as the distances between home and work for Taudion Town employees, we can use the symbol X.[2] (Later, when we deal with two variables at the same time, we use Y to symbolize the other set of observations.

To indicate the operation of *summation,* we use the symbol Σ, which is the uppercase Greek letter sigma. Hence, ΣX instructs us to sum all values of X or, in our example, to sum all the distances between home and work for Taudion Town employees. The use of this symbol for summation is universal.

To symbolize arithmetic mean, we adopt the rather common convention of $\overline{X}$ ("X-bar"), which tells us that we are talking about the mean of the measures on the X variable. With these symbols, we can present the formula for the arithmetic mean:

$$\overline{X} = \frac{\Sigma X}{N} \tag{3.1}$$

This formula, literally translated, instructs us to sum all values of X (that is, ΣX) and divide this summed value by the total number of cases (N).

The Mean as Balance Point.   The arithmetic mean has popular use. Indeed, literate persons everywhere in the world probably use this averaging method. The popularity of the method, unfortunately, is not matched by an understanding of it. Consequently, people often draw incorrect or inadequate conclusions from it and then make wrong decisions.

If the median is the middle value in a set of observations, what does the mean represent? In what sense is it a measure of central tendency? For an answer, let's go to the familiar seesaw. Your experience tells you that, if two persons of equal weight were to sit on a seesaw, they should sit at equal distances from the balance point (if they are to have fun rather than merely aggravate each other). You probably also know that if one is heavier than the other, the heavier sits close to the balance point. (You may recall from high school physics that the **balance point** is found when the distance times the weight on one side equals the distance times the weight on the other.)

The arithmetic mean is such a balance point. Consider these three scores on a test: 5, 6, 10. In Figure 3.1 we put them along a line, thinking of each score as a little box and all boxes having the same weight. The mean of the scores is (5 + 6 +

---

[2] A proper statistician may wince at this. In formal statistics, more precise symbolization is to let X stand for the *variable* (distance between home and work) and X_i (read "X sub i") for the value of a particular observation on the variable. The refinement is not necessary for elementary descriptive purposes.

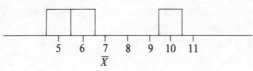

FIGURE 3.1

10)/3 = 7, and in Figure 3.1 the point 7 is the precise balance point for the three scores. When we add the distances between each score and the mean, we find that the distances on the left (2 + 1) equal the distance on the right (3). The mean, then, is the balancing point of a distribution; it is the point from which individual score deviations (distances) in one direction are exactly balanced by deviations (distances) in the opposite direction. If we assign negative values to the deviations to the left of the mean and positive values to the deviations to the right of the mean, we can say that the sum of all deviations about the mean is zero.

You should now understand that the median and mean are quite different concepts of central tendency—both are informative, but they carry different information. It is highly unlikely that, for a given distribution, the mean will exactly equal the median. This will occur only when the distribution is perfectly symmetrical, that is, when the frequencies and distances on one side equal those on the other.

The Mean from Grouped Data

When data are grouped in a frequency distribution, it is not possible literally to sum the individual measures because, as you already fully know by this time, we have lost their identities. Grouping, however, does not paralyze us. We can use the midpoints of the class interval to stand for the values of the observation in the interval. The mean computed from grouped data usually is not precisely what would have been obtained from the original measures but the approximation is fairly good. As the intervals are made coarser, the approximation similarly coarsens. In any event, it is proper to view the mean calculated from grouped data as the mean *of that* particular distribution.

Determining the mean of a frequency distribution is conceptually simple, albeit somewhat tedious. We shall consider three different but equivalent methods and tell you in advance that the third method will be the most convenient of the three. However, you will more easily understand its rationale if we explain the others first.

Computation of the Mean Directly from Midpoints. If the midpoint of an interval represents the value of all scores in that interval, then multiplying the interval midpoint by the interval frequency denotes the sum of scores in the interval. When this multiplication of the frequency times the midpoint is accomplished for all intervals, the summation of all the products is the summation of scores for the distribution. Dividing the sum by N gives us a mean. We can write a formula for this method, using X' (read as "X prime") to symbolize the midpoints (reserving X to mean the original scores):

$$\bar{X} = \frac{\Sigma\,(fX')}{N} \tag{3.2}$$

Let us use this method to find the mean of the distribution in Table 3.2. (Do not try to grasp all of the detail in this table right now; at the moment we are interested only in the first four columns—the remainder is discussed later.) The first column organizes reading rates into intervals with interval widths of seventy-five. The second column identifies the interval midpoints (which we designate by X'). The third column, f, tells us the frequency in each interval. The fourth column, fX', shows the results of this computation. We sum the fX column to find $\Sigma fX'$, which is 93,375. Now we have all of the data we need to apply formula (3.2). The formula and computations for this method are shown beneath Table 3.2, where we learn that the distribution mean is 622.50 words per minute.

Computation of the Mean by Shorter Methods.   In finding the mean by the foregoing method, you may have to work with very large numbers, as we just did. This burden can be reduced by bringing the numbers down to smaller size, thereby also reducing the possibility of computing errors.

TABLE 3.2   Reading Rate, in Words per Minute, of 150 Entering Freshmen at Teufel Tech

Reading rate	X' (midpoints)	f	fX'	d $(X'-AO)$	fd	d' $\left(\frac{X'-AO}{i}\right)$	fd'
875–949	912	1	912	300	300	4	4
800–874	837	19	15903	225	4275	3	57
725–799	762	29	22098	150	4350	2	58
650–724	687	27	18549	75	2025	1	27
575–649	612	21	12852	0	0	0	0
500–574	537	19	10203	−75	−1425	−1	−19
425–499	462	12	5544	−150	−1800	−2	−24
350–424	387	10	3870	−225	−2250	−3	−30
275–349	312	8	2496	−300	−2400	−4	−32
200–275	237	4	948	−375	−1500	−5	−20

$$N = 150 \qquad \Sigma fd = 1575$$
$$\Sigma fX' = 93375 \qquad \Sigma fd' = 21$$

Formula (3.2)	Formula (3.3)	Formula (3.4)
$\bar{X} = \dfrac{\Sigma fX'}{N}$	$\bar{X} = \dfrac{Xfd}{N} + AO$	$\bar{X} = \dfrac{i(\Sigma fd')}{N} + AO$
$= \dfrac{93375}{150}$	$= \dfrac{1575}{150} + 612$	$= \dfrac{75(21)}{150} + 612$
$= 622.50$	$= 10.50 + 612$	$= 10.50 + 612$
	$= 622.50$	$= 622.50$

To make the calculations easier, we take advantage of a property of numbers. Consider, for example, the numbers 8, 10, 12. If we subtract 2 from each of the numbers, we have 6, 8, 10. Notice that the distances between the original numbers are not changed by the subtraction of the constant amount. Since the mean involves distances among values, you may suspect that one can find the mean of the original numbers by working with their reduced counterparts.[3] Your suspicion would be correct. The mean of the original numbers is $(8 + 10 + 12)/3 = 10$. The mean of the reduced numbers is $(6 + 8 + 10)/3 = 8$. The difference between the two means is 2, which is precisely the value of the constant we subtracted from the originals. If we add the constant to our reduced mean, we have 10, the correct mean for the original scores.

Before writing the formula for this method, let us look again at the distribution of reading rates in Table 3.2. The scores (X') are large numbers. We wish to reduce them by some constant amount. Which constant can we use? Any constant will work. However, we want to use a constant that will result in the maximum reduction in the size of scores. Because maximum reduction is produced by a constant at or near the mean, we guess which interval is likely to contain the mean and use its midpoint (X') as our constant. (The outcome will not be affected if we guess incorrectly.)

Let us call our selected midpoint the *arbitrary origin* (AO), which is the constant we will subtract from each X'. Let us call the reduced scores d (they represent the differences between each X' and the arbitrary origin, $d = X' - AO$). We can write the computational formula for this method as

$$\bar{X} = \frac{\Sigma fd}{N} + AO \qquad (3.3)$$

Now we are ready to find the mean reading rate from Table 3.2 with this method. First, we select the midpoint 612 as our arbitrary origin, which is our guess about the approximate interval in which the mean lies.[4] When we subtract the constant

---

[3] These reduced scores are sometimes referred to elsewhere as *coded* scores.

[4] In this guess we cheated a little. Since we already calculated the mean for this distribution earlier, we know where it lies. However (and cross our hearts), we would have picked this value anyway because of something you will learn later—the relative positions of mode, median, and mean in a skewed distribution.

612 from each X', we obtain the reduced scores in column d of the table. Notice that the reduced scores go up (positively) and down (negatively) from the arbitrary origin we selected, at which point the reduced score (d) is 0. Notice, also, that in each direction from the arbitrary origin, the reduced scores move in increments of the interval width, which is seventy-five. Now, perhaps, you can see why the constant is called the arbitrary origin. When using this method, we find d merely by identifying the arbitrary origin as 0 and then moving up and down in units of the interval width. That is, it is not necessary to go through the tedium of subtracting AO from each X' to find d.

When each d is multiplied by its associated frequency (f), we obtain the next column in Table 3.2, labeled fd. Summing this column, we have all the information required by formula (3.3). The computations are shown beneath Table 3.2, where we see that the mean is 622.50, precisely the same as was produced by the earlier method.

The Shortest Method for Computing the Mean.   We now can show you a method that will reduce the computational burden to pint size. Here we utilize another property of numbers. Consider the numbers 4, 12. If each is divided by 4, we have 1, 3. Notice that the distance between the original numbers also is reduced by the same factor. The mean of the original numbers is $(4 + 12)/2 = 8$. The mean of the reduced numbers is $(1 + 3)/2 = 2$. If the reduced mean is *multiplied* by 4, we get 8, the correct mean for the original numbers. This means that the scores in a distribution can be reduced by dividing with a constant, and the mean of the reduced score can be corrected by multiplying with the same constant.[5] In our shortest computational method, we are going to combine the advantages of subtracting an arbitrary origin and dividing by a constant. If an arbitrary origin is subtracted from each midpoint and each resulting d is then divided by the size of the interval width, something very interesting results. We have done this in Table 3.2. Each d was divided by the interval width (75), and the results are shown in the next to the last column of the table. We label these reduced scores d' (read as "d prime"). As you see immediately, these reduced scores are indeed very small. Equally important is that they can be produced merely by showing the arbitrary origin as zero and then moving up and down in units of one, showing negative values below zero. Neither subtraction nor division is necessary to produce column d' for any distribution!

We find the mean of the d' scores by multiplying each d' by its frequency (last column), adding these products, and dividing by N, being careful with the signs. The mean of the d' scores is corrected back to its original value through multiplication by the size of the class interval and addition of the arbitrary origin midpoint. Using i to symbolize the interval width, the computational formula is

$$\bar{X} = \frac{i\,(\Sigma fd')}{N} + \text{AO} \tag{3.4}$$

---

[5] The converse is true also. When scores are in decimals, you can ease the computations by multiplying with a constant and correcting with a final division by the same constant.

The application of this formula to Table 3.2 is detailed below that table, where we see that the mean is 622.50 words per minute—again, the same result as was obtained with the other two formulas.

Open-Ended Intervals.   You learned earlier that, in the case of the median, the presence of open-ended intervals does not preclude computation unless the median lies in one of the open intervals. What about the mean? Here, an open-ended interval is fatal and there is no antidote. When we find the mean from grouped data, we use the midpoint (X') of each interval to represent the interval scores. The midpoints must be known—which is not possible in the case of an interval whose limits are not identified.

Combining Several Means

It is sometimes the case that we know the means for several groups and wish to find the mean for all the groups combined. The unsophisticated impulse is to find the mean of the means, i.e., to sum the means and divide by their number. This would be incorrect, except in the unusual circumstance where all the N's were equal.

Let us look at two sets of scores and their means:

A	B
10	9
12	11
18	
20	
$\bar{X} = 15$	$\bar{X} = 10$

What is the mean of groups A and B combined? As you know, the mean of all six scores is $(10 + 12 + 18 + 20 + 9 + 11)/6$, which equals 13.33. If we had computed the mean of the two means, we would have found it to be 12.5. Why is the latter procedure incorrect? Because it gives the two scores in B the same weight as the four in A. How, then, do we find a combined mean when all we know is the separate means and the N for each? The procedure, which is simple, provides for "weighting" each mean with its N. For two groups, the computational formula is as follows, in which $\bar{X}_c$ stands for "the mean of combined groups" and the subscript after each $\bar{X}$ and N identifies the two groups:

$$\bar{X}_c = \frac{N_1\bar{X}_1 + N_2\bar{X}_2}{N_1 + N_2} \tag{3.5}$$

The formula for more than two groups merely continues adding $N\bar{X}$ terms in the numerator and N's in the denominator for each additional group.

The formula for combining means is derived directly from the basic formula for the arithmetic mean. If you recall your high school algebra, you can see this for yourself. Start with

$$\bar{X} = \frac{\Sigma X}{N}$$

and then multiply both sides of the equation by N, which gives us

$$N\bar{X} = \frac{N\Sigma X}{N}$$

Since the N's in the right-hand expression cancel out, we have

$$N\bar{X} = \Sigma X$$

We see, therefore, that the summation of scores can be determined by multiplying the mean by the number of cases. If you look at formula (3.5) carefully and with this in mind, you can see that the numerator consists of the combined summations of scores and the denominator consists of the combined N's.

When we apply formula (3.5) to our little A and B groups, we find

$$\bar{X}_c = \frac{4(15) + 2(10)}{4 + 2} = \frac{80}{6} = 13.33$$

which we know is correct.

Let us use this new knowledge in a more realistic example. A newspaper item tells us that, in a certain small company, the average salary of the three supervisors is $20,400 and the average for the remaining 120 workers is $12,200. (Unfortunately, and typically, we were not told which average was used, but we will assume it was the arithmetic mean.) What is the average salary of all these employees? Formula (3.5) permits us to find the answer:

$$\bar{X} = \frac{3(20,400) + 120(12,200)}{3 + 120}$$

$$= \frac{61,200 + 1,464,000}{123}$$

$$= \frac{1,525,200}{123}$$

$$= \$12,400$$

Your pencil will tell you that a quite different (and incorrect) answer is given if you simply find the mean of the two mean salaries.

MEASURES OF CENTRAL TENDENCY COMPARED

You have learned about three ways in which the central tendency of a set of measures can be conceptualized and calculated. In deciding which to use, and in interpreting those you are giving, it is important to understand how they differ.

Simplicity of Interpretation

The mode probably is the easiest to interpret and explain. It is the particular observation that occurs most frequently in a set of data. The median, similarly, is easily understood to be the middle value of a set of data: the value dividing the lower and upper halves of the total cases. The mean, despite its more popular use, is the most complex of the three: the point in a distribution that balances out the deviations in both directions.

Sensitivity to Grouping Error

The simplicity of the mode is offset by the fact that its location from grouped data varies greatly with different decisions about interval width and interval limits. This sensitivity to interval width and limits also is characteristic of the median but to a much smaller extent. The mean is the clear winner against this criterion, showing the least sensitivity to grouping decisions.

Sensitivity to Distribution Shape

Look again at Figure 3.1 where we showed three scores, 5, 6, and 10 on a scaled line. The median is 6 and the mean is 7. If the right-hand score (10) were changed to 7, what changes would take place in the two measures of central tendency? The median would be unaffected; 6 would remain as the middle value. The mean, however, would become $(5 + 6 + 7)/3 = 6$. The median, as a counting measure, has no interest in the magnitude of a given score; it is interested only in whether the score is below or above the midpoint. How *far* above or below is irrelevant and ignored. The mean, on the other hand, takes the magnitude of each score into account; a change in any score necessarily results in a change in the mean.

In a skewed distribution, therefore, the mean and median are different values. The deviations from the midpoint in one direction are not compensated by deviations in the other direction. The mean and median are at the same point only in a symmetrical distribution.

The relative positions of mean and median in skewed distributions are predictable. The mean in such distributions is influenced in the direction of skewness. It can be said to "chase the tail" of the distribution. We illustrate this in Figure 3.2 which shows two distributions of different skewness with the relative positions of mean and median in each case. Because the distributions are unimodal, Figure 3.2 also can show the relative position of the mode. Notice that the mean is to the right of the median in a positively skewed distribution, and that the converse is true in a

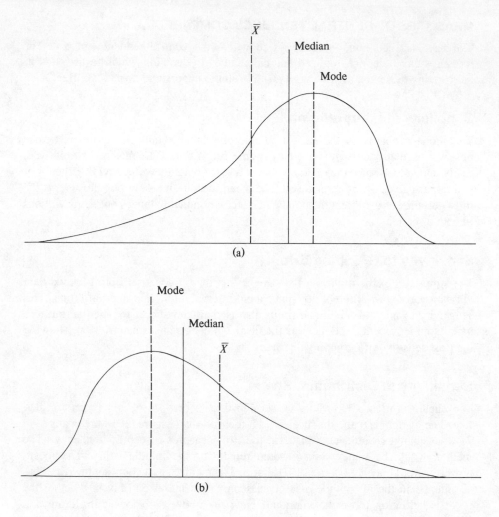

FIGURE 3.2 Relative positions of measures of central tendency in (a) negatively skewed and (b) positively skewed distributions.

negatively skewed distribution—it chases the tail. (If you need a memory aid, it happens that, in a negatively skewed unimodal distribution, the mean, median, and mode fall in alphabetical order from left to right.)

Because of its sensitivity to extreme scores, the mean may be a deceptive measure of central tendency in describing markedly skewed distributions. Take, for example, the annual income of families in a small community in which everyone is impoverished except for one extremely wealthy person. In that unhappy distribution, only one case may fall above the mean; that is, reporting only the mean may give a false impression of what is generally the case. The median, and perhaps the mode, would seem to be more appropriate, would it not?

In Chapter 2, you saw that graphic description of data can be distorted to create different impressions for the audience. Self-interested reporting is possible also by using only the most advantageous average in a particular situation. Hear, for illustration, the following exchange between two workers' groups competing for higher salaries: "The average salary of hedge trimmers," said the President of the Hedge Trimmers Association," is lower than the average salary of sidewalk washers." "Not true," said the President of Sidewalk Washers. "The average salary of sidewalk washers is much lower than that of hedge trimmers." It turns out that both statements were correct, each employing a different definition of average—and on distributions skewed strongly in opposite directions. The situation is shown in Figure 3.3. If the mean is used, hedge trimmers, indeed, have the lower "average." Conversely, the sidewalk washers are in the poorer position if the median is employed.

When describing the central tendency of skewed distributions, the median usually is more appropriate, along with the disclosure that the distribution is skewed. Most analysts report both mean and median, which we think is a more informative habit. A difference between mean and median suggests that the distribution is skewed and the difference indicates also the probable direction of skewness.

The Effect of Grouping Error on Mean and Median Determinations

When we find the median from grouped data, we assume that the scores in an interval are distributed uniformly throughout that interval. In the determination of the mean, we assume that the interval midpoint can represent all the scores in the interval. How good are these assumptions? The answer is that, although the assumptions are necessary to permit calculations, they do not square perfectly with reality. Hence, as we have reminded you, the mean and median from grouped data usually

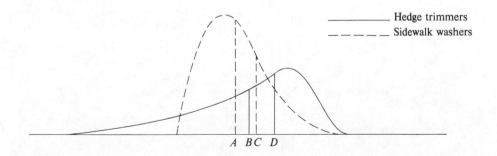

FIGURE 3.3 **Distributions of salaries of hedge trimmers and sidewalk washers. Mean and median for trimmers are at points *B* and *D*, respectively; mean and median for washers are at *C* and *A*, respectively.**

will not be exactly those that would be obtained from the original observation. However, the error usually is tolerable for most practical purposes. Let us see why.

We give you Figure 3.4, which is a perfectly symmetrical frequency distribution. On the base line, we have marked off three intervals: 2.5–4.5, 4.5–6.5, and 6.5–8.5. The midpoints of the intervals are shown by the dotted lines. Consider first the interval to the right of the distribution mode, with interval limits 6.5–8.5. What can you say about the way in which scores are distributed in the interval? If you examine the area of the curve that is marked off by these interval limits, you see that there is more area (that is, there are more scores) below the interval midpoint than above it. The midpoint of this interval therefore does not correctly represent the average of these scores in the interval—it is too high; the mean of actual scores in the interval is a smaller value than the midpoint value (7.5).

So much for the bad news; now for some better news. Examine the interval 2.5–4.5 in Figure 3.4, which is the mirror opposite of the interval we have just considered. In this interval we find more area (more scores) above the interval midpoint than below. Therefore, the midpoint of this interval represents an errone-ously *low* estimation of the central tendency of actual scores included in the inter-val. Furthermore, the low-side error here is of the same order of magnitude as the high-side error that is present in interval 6.5–8.5, the mirror-opposite interval. You probably already sense the conclusion. Before we draw it, however, look at the middle interval in Figure 3.4. It has been placed exactly in the middle of the distribution. If you examine the shape of the curve above the middle interval, you can see that the number of scores above and below the interval midpoint are equal and that the shapes on each side are similar. Therefore, in that interval, the midpoint correctly reflects the central tendency of interval scores.

We therefore can say that, in a unimodal, symmetrical distribution, the determi-nation of central tendency is not greatly affected by grouping error: the error on one side of the distribution tends to be matched by an equal and opposite error on the other side of the distribution. That is, errors on one side are compensated by errors on the other.

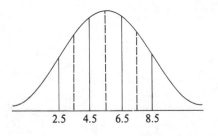

FIGURE 3.4   Illustrations of graphing error in a symmetrical distribution.

Many variables tend to distribute themselves symmetrically; their medians and means therefore can be computed from grouped data with reasonable approximation of the true central tendencies. The situation in markedly skewed distributions is not so comforting, however. Perhaps you can understand from the foregoing discussion that compensating errors are not obtained with skewness (mirror opposites for intervals do not exist). For positively skewed distributions, computation of median and means from grouped data results in erroneously high values. The converse is true for negatively skewed distributions: the values are erroneously low. The greater the degree of skewness, the larger the error due to grouping.

Best Guess Comparison of Mode, Mean, and Median

The three measures of central tendency may be compared also in terms of the basis each affords for guessing about the location of a randomly selected case. The practical value, for descriptive purposes, of this comparison is low but it helps us better to understand the differences among the three averages. If you want to pick the location of a particular score and wish to be *exactly* right most often, the mode is the best bet. It is the most frequent score; if wrong guesses, no matter how far off, are equally bad, the mode is the best guess.

If, on the other hand, it is not important that you guess *exactly* right but wish to minimize the probability of guessing too high or too low, the median is the wise decision. More scores are closer to the median than to any other point on the distribution.

If you wish to guess in such a way that, in the long run, your guessing errors average out to zero, taking both the size and direction of the errors into account, the mean is the best guess. It is the point at which the deviations in both directions add to zero.

In discussing central tendency, we have stayed with the purposes of descriptive statistics, which is the subject of this book. For the purposes of statistical inference (inferring group characteristics from samples drawn from the larger group), the mean has hands-down superiority over the median and mode. It is the most stable measure across different random samples drawn from the same group or population. Furthermore, of the three averages, it is the only one that is algebraic; as a proper member of the mathematics family, it can enter into advanced analyses.

EXERCISES

1. Jane Doe is a precocious twelve-year old whose $2.00 weekly allowance does not fit her preferred life style. In order to prepare a good argument for her parents, she checks among her six best friends and finds that five receive the same as she ($2.00) and the lucky other gets $4.00. She analyzes her data, searching for a self-favoring but truthful way to finish this proposed announcement to her parents: "Compared with the allowances of my best friends, mine is _____ ." Fill in the blank for her.

2. Examine the distribution shown below and answer the questions that follow, without performing any computations:

X	f
More than 29	1
25–29	3
20–24	4
15–19	10
10–14	18
Less than 10	22

   a. What is the general shape of the distribution?

   b. In which interval does the median lie?

   c. What is the *crude mode?*

   d. What are the real limits of the interval 20–24?

   e. What are the real limits of the top interval?

   f. In what interval does the mean lie?

   g. Which is probably greater, the mean or median?

3. For a set of 100 scores, the median is 50. Suppose that we removed the scores 45 and 60. Would the median be affected?

4. Return to the situation in the preceding problem. The mean of the 100 scores also is 50. What will the new mean be (after dropping the two scores)? Hint: What is the original ΣX, what is the new ΣX, and what is the new N?

5.   What are the mean, median, and mode for the following scores?

$$2, 3, 3, 4, 4, 5, 8, 8, 8, 9, 12$$

6. A sociologist studying family size in five small urban towns found the following:

Town	Mean family size	Number of families
A	3.5	802
B	5.4	320
C	4.0	524
D	3.8	731
E	4.5	410

What is the mean family size in all five towns combined?

7. You see below a frequency distribution of the prices of 512 houses sold in a California community in 1976:

Price of house	f
160,000–179,000	4
140,000–159,000	8
120,000–139,000	18
100,000–119,000	25
80,000– 99,000	41
60,000– 79,000	100
40,000– 59,000	206
20,000– 39,000	110

a. What is the interval width?

b. What are the real limits of the bottom interval?

c. How would you describe the shape of the distribution?

d. Which do you predict is larger, the mean or the median?

e. What is the median price of houses?

f. Use formula (3.4) to find the mean.

8. During a one-hour period, the speed of cars passing a highway checkpoint was recorded electronically. Here are the grouped results:

Speed (mph)	f
80–84	3
75–79	9
70–74	17
65–69	21
60–64	30
55–59	37
50–54	17
45–49	8
40–44	5
35–39	2
30–34	1

(Problem continues on next page.)

   a. What are the modal and median speeds?

   b. Using formula (3.2), find the mean speed.

9. The following grouped distribution displays the Stanford-Binet IQ's of 170 youngsters attending a summer camp. Find the median and mean, using formula (3.3) for the latter.

IQ's	f
146–150	2
141–145	1
136–140	4
131–135	7
126–130	10
121–125	11
116–120	14
111–115	18
106–110	24
101–105	25
96–100	16
91– 95	16
86– 90	8
81– 85	6
76– 80	5
71– 75	3

10. An investigator, after computing the mean, median, and mode for a set of scores, learned that the highest score was slightly in error. What effect will the correction have on the measures of central tendency?

11. Make a guess about the shape of distributions in each of the following situations:

	Mean	*Median*	*Mode*
a.	12	16	17
b.	14	14	14
c.	20	15	13
d.	15	15	10 and 20

12.   Assuming that the first statement in the following advertisement is true, what might be

deceptive about the second? "More people fly on Rinkydink Airlines than on any other airline. Come and join the crowd."

13. Suppose that your retirement income is based on the average of the salaries of the last five years of employment. If each year's salary is higher than the preceding, which definition of average would you prefer? Why?

14. If you wished to improve the *median* performance of a group of people (and nothing else mattered), which of the following subgroups would get most of your attention? (a) very low performers; (b) those slightly below median performance; (c) those slightly above the median; (d) very high performers. Suppose you wished only to improve *mean* performance; which group would get your particular attention?

15. Using the grouped data presented in Table 2.4, on p. 14, find the $\bar{X}$, the median, and the mode.

4

DESCRIBING VARIABILITY

You have learned how to describe the central tendency of a set of data by the use of the statistical simplifiers: mode, median, and mean. Central tendency, however, is only part of the descriptive picture. A full representation of a collection of observations also should include some statement about its scatter or variation. Without an indication of the variability present in a distribution, central tendency is not sufficiently informative and may even be misleading. You would be unwise, for example, if you packed your bag for an October trip to Midcity knowing only that the mean October temperature is 68°F. Surely you would want to know how much fluctuation you might expect in daily temperatures. Similarly, Tom, a teacher, cannot make intelligent selection of reading materials for his new class if he is told only the mean reading level of the incoming pupils; he must know also something about how much these pupils vary in reading levels. Mary, a merchandiser, cannot plan the most effective sales campaign for her region if she is informed only that the median family income is $14,000.

There is an unhappy story, oft-told and probably untrue, of the ancient Chinese general who, leading his army, arrived at a river that had to be crossed. Seeing no bridges and having no boats, he inquired about the depth of the river and was told, "the average depth is only 2 feet." Whereupon, he confidently ordered his army to walk across. Most of them drowned.

Descriptions of central tendency unaccompanied by descriptions of variation or dispersion, therefore, do not tell us enough. Furthermore, descriptors of variation are needed by definition when variability itself is the object of attention—for example, when we wish to find out how two groups compare in variability or when we wish to determine whether the variability in a group changes over time. As you will see in Chapter 5, measures of variation also are useful in assigning scores to individuals.

There are several statistical descriptors of variation. Here, we deal with the *absolute range,* the *interquartile range,* the *mean deviation,* and the *standard deviation.* These terms may seem esoteric at the moment, but by the end of the chapter they will have become your good friends.

ABSOLUTE RANGE

The simplest way to describe the spread of observations in a distribution is to identify the lowest and highest values and note the difference between them. This is called the **absolute range.** Consider, as an illustration, the following scores:

$$9, 11, 11, 12, 14, 16, 20$$

We can describe the spread by noting that the range is 9 to 20—or 11. A slightly different common practice is to report the range of these scores as 9 *through* 20—or 12. In the first method, the range is expressed as the highest value minus the lowest; in the second, the range is found by subtracting the lowest value from the highest plus one. Either practice is acceptable.

Identification of the lowest and highest values in a set of observations is usually useful information and we recommend the practice. However, absolute range as the *sole* descriptor of variation has a serious limitation in that it communicates nothing about how the observations are distributed between the two extreme values. In Figure 4.1, you see two distributions (in the form of histograms) with exactly the same range (10 to 50). However, your eyes should tell you that the individual scores in distribution (b) scatter more than those in (a). The A measurements are more alike than those in (b). If you described both only in terms of absolute range, your audience might draw the erroneous conclusion that the distributions were similar in variation.

There is another difficulty with absolute range as the only descriptor of variation: it depends solely upon the location of the two extreme values in the distribution. Any error in either score changes the range, and it is the case that extreme scores

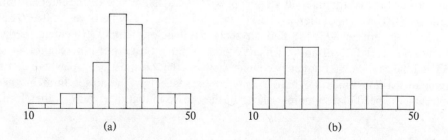

FIGURE 4.1   Two distributions with same range but different dispersion.

tend to be more error prone than more typical scores. For both of these reasons, the more frequent practice is to indicate the range between two values that lie closer to the center of the distribution. The *interquartile range,* to which we now turn, is such a procedure.

THE INTERQUARTILE RANGE

We wish to disregard extreme values and determine the range between two intermediate values. How many of the extreme scores shall we eliminate? We can, as is sometimes done, ignore the bottom 2 percent and the top 2 percent of the scores, thereby identifying the range of the middle 96 percent of the distribution. Or we can select any other percent for elimination as long as we are confident that extremes are eliminated and we do not approach the center of the distribution so closely as to make the descriptor meaningless. A common approach is to ignore the top and bottom 25 percent of the scores, thereby identifying the range of the middle 50 percent of the cases. This descriptor is called the **interquartile range** because, in statistics, the point below which 25 percent of the cases fall is called the *first quartile* (Q_1) and the point below which 75 percent fall is called the *third quartile* (Q_3).[1] Hence, using IR to symbolize *interquartile range,* the formula can be written as

$$IR = Q_3 - Q_1$$

Let us apply the formula to the grouped data shown in Table 4.1. We need to find Q_1 and Q_3. Q_1, we said, is the point on the distribution below which 25 percent of the cases fall. We find that point precisely in the way we found the median in Chapter 3 except that we are looking for 25 percent of the cases (Q_1) instead of 50 percent (the median).

In our illustration, $N = 44$, 25 percent of which is 11. Q_1, therefore, is the point below which 11 cases fall. We examine the cumulative frequency column and note that 9 cases fall below 14.5 (the upper real limit of the score interval 10–14). We need two more cases: two of the 13 that are in the next interval. Using the assumption and logic for finding the median earlier, we determine that Q_1 is 14.5 (the lower limit of the interval 15–19) plus 2/13 times the size of the interval (the latter is 5). The computation is worked out below Table 4.1, where we see that Q_1 is 15.3. To determine $Q3$, we find the score point below which 75 percent of the forty-four cases fall (or above which 25 percent fall). Seventy-five percent of 44 is 33 and the cumulative frequency column shows us that Q_3 must be somewhere in the interval 20–24. Since we have twenty-two cases below the lower limit of that interval, we need eleven more—eleven of the twelve in that interval. Q_3, therefore,

---

[1] The second quartile point (below which half the cases fall) is, of course, the median. A common error in language surrounds the term *quartile*. A quartile is a *point,* not an area. The first quartile (Q_1) is the point below which the bottom quarter of the cases falls. It is correct to say that a score falls below the bottom quartile; it is incorrect to say that it falls *in* the bottom quartile.

TABLE 4.1 Illustration of the Calculation of the Interquartile Range

Scores	f	Cum. f
30–34	2	44
25–29	8	42
20–24	12	34
15–19	13	22
10–14	7	9
5–9	2	2
	$N = 44$	

$Q_1 = 14.5 + (2/13)(5) = 14.5 + .8 = 15.3$

$Q_3 = 19.5 + (11/12)(5) = 19.5 + 4.6 = 24.1$

$\text{IR} = Q_3 - Q_1$

$\quad = 24.1 - 15.3$

$\quad = 8.8$

is 19.5 (the lower real limit) plus 11/12 times the interval width. Working through the computations, we find that $Q_3 = 24.1$. The interquartile range, $Q_3 - Q_1$, is 8.8. This means that the middle 50 percent of the cases in our distribution range over 8.8 score points. This is a relatively simple way of describing variation in a set of observations and is interpretable no matter what the distribution shape may be. Its advantage over absolute range is that it is not influenced by the atypicality of extreme scores.   You may run across a measure of variation called the **semi-interquartile range,** symbolized by Q. This, as its name implies, is one-half of the interquartile range, that is, $Q = (Q_3 - Q_1)/2$. We make no particular point of this procedure here; it usually is not more informative than IR.

There is an inherent weakness in using the range of scores to describe their variability. It depends upon the location of only two scores in the distribution. Consequently, range, whether absolute or some refinement such as IR, ignores the pattern of variation represented among the remaining scores in the distribution. (For this reason, incidentally, range is not used in statistical inference; it is very unstable across different samples of the same population.) We would be better off with a description of variation that reflects the variability of *all* scores in a distribu-tion. You should not conclude, on the other hand, that range is not a useful index of dispersion. We suggest, indeed, that it be reported because it gives some informa-tion about a distribution that is easily interpreted and frequently useful. The proper conclusion is that range should not be the *only* descriptor of variability.

THE MEAN DEVIATION

The general approach to describing the variation present among *all* scores is to identify how much each score in a distribution varies from some common reference point. What reference point shall we use? By this time, you probably anticipate that a reference point in the middle of the distribution is more solid than a point toward either extreme. The **mean deviation**[2] (MD) is a procedure for describing the variation of scores from either the median or the mean. This procedure is rarely, if ever, used at the present time, for reasons given later. However, it is of historical interest and more important it has instructional value in introducing a more popular and sophisticated procedure. Before considering the formula for the mean deviation, let us work through the logic of the procedure, using the mean as our reference point. Consider the following scores:

$$2, 4, 5, 7, 12$$

The mean of the five scores is 6, a statement you can verify. Let us determine how far each score deviates from the mean, *ignoring the direction of the deviation.* Working from left to right, we find that the deviations of the five scores from the mean of 6 are as follows:

$$4, 2, 1, 1, 6$$

If we now find the mean of these deviations, we have a measure of the variation in the set of scores—the mean deviation (from the mean), which works out to be 2.8. The same procedure can be applied with the median as our reference point. The median score is 5 (we hope you can agree) and the deviations of the scores from this median are:

$$3, 1, 0, 2, 7$$

The mean of these deviations is 2.6, which is the mean deviation (from the median). Although either the mean or median can be used as reference point, the mean is more customary.

The formula for the mean deviation utilizes two new symbols. We use x to stand for the deviation of any score from the mean; that is, $x = X - \overline{X}$. We also need a way of showing that the direction of deviation, whether positive or negative, is ignored. The standard convention in mathematics and algebra is to use vertical parallel lines on each side of a value when sign is ignored and only the absolute value is considered. With these understandings, the formula for the mean deviation (whether from the mean or the median) can be written as

$$\text{MD} = \frac{\Sigma \, |x|}{N} \tag{4.1}$$

The right-hand side of the formula tells us to sum the absolute values of the deviations from the mean and to divide the result by the number of scores.

---

[2] Sometimes referred to as the *average deviation.*

The mean deviation clearly is far from complex in concept or computation. Why, then, is it not used in contemporary descriptive statistics? The reason is that, in ignoring signs, we commit a secular sin. The mean deviation, because it overlooks the signs of the deviations, has no algebraic properties; it cannot enter into further statistical analyses or other formulas. We require a descriptor of variation of scores from the central tendency that takes the direction (sign) of the deviations into account without violating any rules of number and algebra. Fortunately, there is such a procedure so we need not draw this chapter to a close at this point.

THE STANDARD DEVIATION

Let us consider the same five scores that we used above. They were 2, 4, 5, 7, 12, and their mean is 6. The *signed* deviations of these scores from their mean are (from left to right): $-4, -2, -1, 1, 6$. That is, the first three scores are below the mean and therefore have negative values. If we add these deviations, attending to the signs, we get a big, fat zero—and you should not be surprised. By definition, you should remember, the sum of deviations about the mean is zero; the mean balances out the deviations on each side. Accordingly, the mean of the signed deviations also is zero—and will be zero every time, in any distribution. Hence, the mean of the signed deviations cannot be used as a measure of variation; it always will be zero.

We need a way of getting rid of those troublesome signs properly. Our salvation is found in the rule that the product of two numbers with like signs is positive. Therefore, if we multiply each deviation by itself (square it), the products will all be positive. For example, $-4(-4)= 16$. We show below the original five scores (X), the deviation of each from the mean (x), and the squared deviations from the mean (x^2):

X	2	4	5	7	12
x	-4	-2	-1	1	6
x^2	16	4	1	1	36

As you can see in the last row (x^2), we now have described the deviation of each score from the mean, all with a common sign (positive) and without violating any number rules. Furthermore, the sum of these squared deviations is not zero and, in fact, will increase in size as the deviations from the mean increase.

The mean of the squared deviations from the mean, therefore, is a measure of variation. In statistics, this mean of the squared deviations is called **variance** and its formula is written as

$$\text{variance} = \frac{\Sigma x^2}{N} \tag{4.2}$$

which instructs us to square each deviation from the mean (x^2), sum the products

(Σx^2), and divide by N. Applying formula (4.2) to our five scores, we have

$$\text{variance} = \frac{16 + 4 + 1 + 1 + 36}{5}$$

$$= \frac{58}{5}$$

$$= 11.6$$

Some prefer or find it convenient to compute variance without finding the deviation of each score from the mean. One of several formulas, derived[3] from formula 4.2, that do not involve x, is the following:

$$\text{variance} = \frac{N\Sigma X^2 - (\Sigma X)^2}{N^2} \qquad (4.3)[4]$$

Do not confuse ΣX^2 and $(\Sigma X)^2$; the ΣX^2 indicates that we square each score and add the squares; $(\Sigma X)^2$ tells us to add the scores and then square the sum. Formula (4.3) applied to our five scores should and does give us the same result as above:

$$\text{variance} = \frac{5(238) - (30)^2}{5(5)}$$

$$= \frac{1190 - 900}{25}$$

$$= \frac{290}{25}$$

$$= 11.6$$

Variance is a central and basic measure of variation in statistical analysis and looms particularly important in statistical inference. Although it is appropriate also in descriptive analyses and presentations, it presents a difficulty in interpretation. Specifically, when we square the deviations from the mean, we end up with values that no longer are on the scale of scores. That is, the variance of a distribution of measurements cannot be interpreted in terms of the units of measurement. However, variance can be brought back to the units of the measurement scale by taking its

---

[3]Since this book is written for those whose algebra is gone or shaky, we do not show the derivation of this or other formulas. If you are comfortable with algebra, however, you need not take us on faith. Merely expand the basic formula $\dfrac{\Sigma(X - \bar{X})^2}{N}$ to derive a number of different computational formulas. If you do so, you will encounter the expression $\Sigma \bar{X}$, which may puzzle you unless you know that it is equivalent to $N\bar{X}$.

[4]You usually will see the formula for variance with $N - 1$ in the denominator, whereas we show N. Our formula is correct for determining the variance of the group of data at hand. When the task is one of *inference* (estimating population values from sample values), $N - 1$ is more appropriate, for reasons too complex to engage here. Be assured that the formula is correct as long as your purpose is descriptive rather than inferential.

square root. If the variance of a distribution of IQ's is 225, we cannot say that it represents 225 IQ units. However, its square root (15) is expressed in IQ units.

The square root of variance is called the **standard deviation,**[5] which we symbolize as s. Its formula, of course, is

$$s = \sqrt{\frac{\Sigma x^2}{N}}$$

or, an equivalent in terms of the actual scores,

$$s = \sqrt{\frac{N \Sigma X^2 - (\Sigma X)^2}{N^2}} \qquad (4.5)$$

The Standard Deviation from Grouped Data

The formulas we have just given you ((4.4) and (4.5)) refer to the original (that is, ungrouped) scores. We use some modification of these formulas when computing the standard deviation from a frequency distribution in which the scores have been grouped into intervals, because the original scores are unknown. Let us take Table 4.2 as an illustration and exercise. There, we repeat the distribution of reading rates presented in the previous chapter. The first column shows the apparent intervals, the interval midpoints (X') are indicated in the second column, and the interval frequencies (f) are given in the third column. (Ignore the remaining columns for the moment.) We could find the standard deviation of this distribution directly from the midpoints, could we not? Applying formula (4.5) and letting X' represent all scores in the interval, we can restate the formula to read as follows:

$$s = \sqrt{\frac{N \Sigma f X'^2 - (\Sigma f X')^2}{N^2}} \qquad (4.6)$$

Consider, however, that applying this formula to our data would require working with fantastically large numbers—an unbearably tedious and error-inviting task if you are computing by hand. Therefore, we ignore this computing approach. Instead, we will reduce the size of the computation work by the same procedure you learned in Chapter 3. That is, we change each X' into the much smaller d' shown in the fourth column of Table 4.2. This change, you remember, represents the subtraction of an arbitrary origin from each X' and division by the width of the class interval (i). Actually, of course, all we need do is arbitrarily select an interval somewhere in the middle of the distribution, set $d' = 0$ at that point, and number the intervals sequentially up (positive) and down (negative) from that arbitrary origin. Now, with

---

[5] We use s here, but you frequently will find the following symbols for the standard deviation: SD, S, and the lowercase Greek sigma, σ. The σ conventionally is reserved to represent the standard deviation of a population (as distinguished from a sample) in statistical inference.

TABLE 4.2 Computation of Standard Deviation for Grouped Reading Rates Given in Table 4.2

Reading Rates	X'	f	d'	fd'	d'^2	fd'^2
875–949	912	1	4	4	16	16
800–874	837	19	3	57	9	171
725–799	762	29	2	58	4	116
650–724	687	27	1	27	1	27
575–649	612	21	0	0	0	0
500–574	537	19	−1	−19	1	19
425–499	462	12	−2	−24	4	48
350–424	387	10	−3	−30	9	90
275–349	312	8	−4	−32	16	128
200–274	237	4	−5	−20	25	100
		$N = 150$		$\Sigma fd' = 21$		$\Sigma fd'^2 = 715$

Formula (4.7)

$$s = i \sqrt{\frac{N\Sigma fd'^2 - (\Sigma fd')^2}{N^2}}$$

$$= 75 \sqrt{\frac{(150)(715) - 441}{22500}}$$

$$= 75 \sqrt{\frac{106809}{22500}}$$

$$= 75(2.18)$$

$$= 163.41$$

d' standing for the interval scores, we can find their standard deviation by a formula equivalent to formula (4.5) but restated in terms of d':

$$s \text{ (of } d') = \sqrt{\frac{N\Sigma fd'^2 - (\Sigma fd')^2}{N^2}}$$

This formula, however, gives us the standard deviation in terms of d', whereas we wish to have the standard deviation in terms of reading rate scores. The conversion is simple; we merely multiply by the width of the class interval (i), thereby reversing the earlier division step.[6] Therefore, the full formula for finding the standard deviation from grouped data, using this short method is

$$s = i \sqrt{\frac{N \Sigma fd'^2 - (\Sigma fd')^2}{N^2}} \qquad (4.7)$$

We are now ready to find the standard deviation of the distribution of reading rates given in Table 4.2, using formula (4.7). The required information is worked out in the last three columns. The fifth column indicates the frequency times d' for each interval; the sixth column shows the square of each d', and the last column shows the frequency times d'^2 for each interval. The summations needed for formula (4.7) are given at the bottom of the fifth and last column. We need also to have the width of the intervals, which is 75, and N, which is 150. Substituting these values in formula (4.7), we have the computations that are worked out beneath Table 4.2. The standard deviation of the distribution of reading rates turns out to be 163.41.

The standard deviation has two general uses in descriptive statistics:

1. The comparison of the variations of two or more sets of data.

2. Interpretation of the variation in a particular set of data using a theoretical model called *the normal distribution*.

We discuss each use in turn.

Comparison of Variations

Where two or more groups have been measured on the same scale, their variability can be compared directly by comparing the standard deviations. Suppose, to illustrate, we have scholastic aptitude test scores for entering freshmen in three colleges and we find their means and standard deviations to be the following:

$$\text{College A:} \quad \bar{X} = 550, \ s = 120$$
$$\text{College B:} \quad \bar{X} = 555, \ s = 125$$
$$\text{College C:} \quad \bar{X} = 550, \ s = \ 90$$

We see that the three groups do not differ very much in central tendency but there are differences in the dispersions. The freshmen at Colleges A ($s = 120$) and B ($s = 125$) are more heterogeneous in scholastic aptitude than are their counterparts at College C, where the standard deviation is much smaller ($s = 90$).

[6] If your memory is good, you will recall that, in finding the mean with an arbitrary origin, we also had to add back the arbitrary origin. In the case of the standard deviation, this is not necessary because we are interested only in the deviations from the mean, which are unaffected by the subtraction of the arbitrary origin.

Standard deviations also permit the comparison of the variability of the same group at different times—so long as the same measurement scale is involved. An example of such a comparison is found in the administration of a spelling test to a group of students before and after special spelling drill. Before drill, we find $\overline{X} = 45$, $s = 10$; after the drill sessions, we find $\overline{X} = 60$, $s = 15$. These statistics tell us that the group improved (the second mean is larger than the first) and that it became more heterogeneous on the measure after drilling (the second s is larger than the first).

Standard deviations for different sets of scores, then, can be compared directly—differences between standard deviations reflect differences in dispersion if the same measurement scale is involved. It is sometimes useful or necessary to compare the variations of distributions based on different measurement scales; that is, how can we compare apples and oranges? An answer is found in the *coefficient of variation*.

The Coefficient of Variation (CV).   Suppose that the mean weight of a group of people is 130 pounds with $s = 13$ and that their mean height is 68 inches with $s = 17$. Is this group more variable in height or in weight? Since each s is expressed in different units of measurement (one in pounds and the other in inches), they cannot be compared directly. However, we can express each relative to its own mean. More specifically, we can express the standard deviation as a percent of the mean. This percent is called the **coefficient of variation.**[7] This is done by dividing s by $\overline{X}$ and multiplying by 100—the multiplication converts the decimal proportion into a percent. The formula is

$$CV = \frac{s}{\overline{X}} \cdot \left(100\right) \tag{4.8}$$

Returning to our example, we can compute two coefficients of variation, one for weight and one for height:

$$CV \text{ (weight)} = (13/130)100 = 10$$
$$CV \text{ (height)} = (17/68)100 = 25$$

The two coefficients of variation can be compared directly. They tell us that the group is relatively more heterogeneous in height than in weight. The coefficient of variation is independent of the units of measurement and therefore is useful in comparisons across different units of measurement. In fact, it also is applied appropriately when comparing two sets of scores derived from the same measurement but representing very different means. If we were to compare the weights of a group of adults with the weights of a group of infants, we would find the adult mean and standard deviation to be much greater than those for the infants. A comparison of the standard deviations would lead to the conclusion that adults are more

---

[7] Sometimes called the *coefficient of relative variation* or the *relative standard deviation*.

heterogeneous in weight. A comparison of the coefficients of variation, however, might reveal that infants are more variable in weight, relative to their mean weight, than adults.

The Standard Deviation
and the Normal Distribution

We have seen that the standard deviation permits comparison of the variations present in two or more sets of observations. The set having the larger s or coefficient of variation can be said to have the greater dispersion or spread. Under a very special circumstance, it is possible also to interpret the variability in a single set of observations through the standard deviation. That circumstance is present when the distribution at hand resembles a particular theoretical distribution called the **normal distribution** or the **normal curve**.[8] The normal distribution is a theoretical invention, an idealized conception of the form a distribution might take under certain assumptions. As is always the case, the real world does not conform precisely to the idealized theory or model. The actual scores or other observations you collect will not represent exactly a normal distribution. However, real distributions sometimes approximate the normal curve sufficiently to permit its use in descriptive statistics. (In the domain of statistical inference, the normal distribution is a central and valuable concept; indeed, it can be called the work horse in making inferences about populations from sample data.)

The theory and properties of the normal distribution are beyond the scope of this book, requiring much more mathematical sophistication than we are assuming among readers. For our purposes, it may suffice to note that the normal distribution has the following properties:

1. The normal distribution is unimodal and perfectly symmetrical around its own mean. Its characteristic shape is that of a bell and, therefore, is sometimes called *the bell-shaped distribution*.[9] Since it is unimodal and symmetrical, the mean, median, and mode are exactly the same.

2. In the normal distribution, there is a precise relationship between areas under the curve and units of the standard deviation. (We will unpuzzle this statement immediately below!)

What is meant, in the second property, by "units of the standard deviation"? The best explanation, perhaps, is through an illustration: Suppose we have a distribution of Stanford-Binet IQ's, with $\bar{X} = 100$ and $s = 16$. We can express any IQ value by

---

[8] Developed by the nineteenth century mathematician, Karl Friedrich Gauss; hence, it sometimes is referred to as the **Gaussian distribution** or the **Gaussian curve**.

[9] This label sometimes is used carelessly. As you easily can infer from the next paragraph, all bell-shaped distributions are not necessarily normal.

indicating how many standard deviations it is above or below the mean. For instance, an IQ of 116 can be said to be one standard deviation ($s = 16$) above the mean. Similarly, an IQ of 84 lies one standard deviation below the mean. We use the symbol z to label a score value that is expressed in terms of the number of standard deviations from the mean represented by that score value—adding a $-$ or $+$ sign to show the direction from the mean. In our illustration, an IQ of 100, since it lies exactly at the mean, would have a z value of 0.00. The foregoing can be expressed in a formula for converting any score value to a z value:

$$z = \frac{X - \bar{X}}{s} \tag{4.9}$$

which will yield a negative value for z when the score is below the mean.

We are now ready to explain the relationship between units of the standard deviation and areas under the normal distribution. Look at Figure 4.2,[10] which is a normal distribution. On the base line, we have marked off some values of z, limiting ourselves to five points for our illustration. At each of the selected z points, we have erected a perpendicular (called an *ordinate*). Between the perpendiculars, we indicate the percent of the total area that is included. As you can see, the area between the mean ($z = 0.0$) and $+ 1.0$ z is approximately 34 percent of the total distribution. Or, we can say that 34 percent of all observations in this distribution lie between the mean and a point one standard deviation above the mean. The figure also tells us that approximately 14 percent (13.59 percent) of the area or cases fall between $z = 1.0$ and $z = 2.0$. Because the distribution is symmetrical, the respective areas on the left side of the distribution are the same as those on the right.

You can verify from Figure 4.2 that the middle 68 percent of the cases in a normal distribution lie between -1.0 and $+1.0$. If we go up and down two standard deviations, we see that 95 percent of the total area is accounted for.

Where we have data that approximate a normal distribution, therefore, we can use the standard deviation to offer a description of the variability of the data. For an example, let us return to our Stanford-Binet IQ's (which tend to distribute normally in the general population). If the mean IQ is 100 and the standard deviation is 16, we can say that approximately 68 percent of the population has IQ's between 84 (that is, $-1.0z$) and 116 (that is, $+1.0z$). Any number of other descriptive approximations are possible, of course. Can you not, from Figure 4.2, estimate the percent of IQ's that are greater than 132 (that is, greater than $z = +2.0$)?

In Figure 4.2, you are given information on a few z points. However, theoretical proportional areas are determinable for any value of z and are available in tables of varying detail. Appendix II permits you to make approximations of the proportional areas under the normal curve marked off by all possible z values from .00 to 3.70. A little practice will enable you to read the Appendix II table easily. The first column lists the z values, the second column tells you the proportional area between the mean and each z value, in either direction. In the third column, you find the

---

[10] You will observe that the two ends of this distribution curve do not come down to the base line. This is not an oversight. The normal distribution theoretically stretches to infinity in each direction—a distance you will agree is awkwardly long!

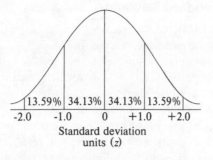

| 13.59% | 34.13% | 34.13% | 13.59% |
| -2.0 | -1.0 | 0 +1.0 | +2.0 |

Standard deviation
units (z)

FIGURE 4.2   Illustration of the relationship between units of standard deviation (z) and areas under the normal curve.

proportional area lying beyond the value of z. As an illustrative exercise, let us find the proportional area lying between the mean and $z = +1.55$. We look down the first column (z) to find 1.55, next to which (in the second column) we find the entry to be .4394. This means that, in a normal distribution, approximately .44 (or 44 percent) of the total area (or cases) fall between the mean and a score that is 1.55 standard deviation above the mean. If we look at the third column entry for $z = 1.55$, we find .0606, telling us that .06 (or 6 percent) of the cases fall above that z value. (For any z value, the second and third column entries add up to .50.)

If we were interested in $z = -1.55$, we would enter the Appendix II exactly in the same way and obtain the same values. However, with $z = -1.55$, we are interested in the left side of the normal distribution; therefore, the third column entry (.060), tells us that 6 percent of the cases lie *below z* = −1.55.

It is important to understand that Appendix II represents only one-half of the normal distribution—either half. Accordingly, if we wish to estimate the proportion of area falling below $z = +1.0$, we first find in the second column that .34 of the area lies between $\bar{X}$ and +1.0 and then add .50 (the proportional area below the mean). The answer is that 84 percent (.34 + .50) of all cases in a normal distribution lie below a score value that is one standard deviation above the mean.

When N is large, the scores on many variables tend toward normal distribution. Educators particularly find this theoretical distribution applicable when interpreting variations against norms for standardized tests, which frequently represent approximate normality of distribution. On the other hand, it is unrealistic to expect or assume even approximately normal distributions in the data you gather and particularly so when N is small. Despite the attention given here to this normal distribution, we counsel against its use in descriptive statistics, unless N is very large and you have been informed that the assumption of normality is appropriate. We especially caution against the frequent error of assuming that a distribution is normal if it is bell-shaped. As was noted earlier, while the normal distribution is bell-shaped, the converse is not true. The relationship between z values and areas tabled in Appendix II apply only to the normal distribution.

Grouping Error and the Standard Deviation

In Chapter 3, you learned that, when the distribution is symmetrical, grouping errors tend to cancel out in finding the mean and median. In Figure 3.4, you saw that the use of the interval midpoint to represent interval scores results in too large a value in intervals above the mean. However, you will recall, interval midpoints below the mean result in too small a value. Therefore, the errors below and above the mean compensate each other in a symmetrical distribution.

This compensation or balancing of grouping errors does not occur in the case of the standard deviation. The reason is easily apparent when you realize that we are considering *deviations from the mean*. In any interval above the mean, the deviation of the interval midpoint is farther from the distribution mean than the correct mean of interval scores would be; that is, the midpoint is an erroneously large statement of the deviations of interval scores from the central tendency. For any interval below the mean, the midpoint is similarly farther from the distribution mean than really is the case for actual scores on this interval; that is, the midpoint also is erroneously large. No compensation is present and, indeed, error is added on error. The result is that the standard deviation computed from grouped data is larger than the true value (the value produced by computation from the original, ungrouped scores).

The error caused by grouping in computing the standard deviation decreases as the number of intervals used becomes larger. Or, to state the relationship in another way, grouping error decreases as the interval width (i) decreases. In fact, if you employ ten or more intervals in grouping your data, the error due to grouping usually is so small that you can overlook it.

Where only a few intervals have been employed, or when you wish a more exact estimate of the standard deviation, you may apply a correction formula called *Sheppard's correction for coarse grouping*. The formula is the following, in which s is the standard deviation computed from grouped data and i is the width of the interval:

$$s \text{ (corrected)} = \sqrt{s^2 - \frac{i^2}{12}} \qquad (4.10)$$

To illustrate the application of Sheppard's correction, let us assume that we have computed $s = 2.50$ from grouped scores and that the interval width (i) is 3. Entering those values in formula (4.10), we have

$$s \text{ (corrected)} = \sqrt{(2.50)^2 - \frac{(3)^2}{12}}$$

$$= \sqrt{6.25 - .75}$$

$$= \sqrt{5.50}$$

$$= 2.35$$

As you can see, the corrected value of s is smaller than the s computed from grouped scores.

There is a limitation on the use of Sheppard's correction: it assumes that the distribution is unimodal and symmetrical. When your distribution departs markedly from either of those two characteristics, the correction should not be applied.

EXERCISES

1. Fifty students in a typewriting course were tested at midsemester and again at the end of the semester. Each student was scored on both occasions in terms of the number of words per minute correctly typed. The results are shown in the following table:

Words per Minute	Midsemester	Semester End
105–109		1
100–104		1
95– 99		2
90– 94		2
85– 89	1	3
80– 84	1	4
75– 79	2	6
70– 74	3	7
65– 69	4	7
60– 64	7	6
55– 59	9	4
50– 54	9	3
45– 49	6	1
40– 44	4	1
35– 39	2	1
30– 34	1	1
25– 29	1	

If you compute the means for the midsemester and semester end scores, you will find that, "on the average," the students improved in typing speed. No surprise, of course. However, what happened to individual differences? As the group got better, did its variability in typing speed increase, decrease, or remain fairly constant? To answer this question, describe the variation in each test in terms of range (R), interquartile range (IR), and standard deviation (s). Use each of these three measures of variation to compare the two test results. Which do you conclude?

2. Over the past thirty days you weighed yourself each morning, keeping a record of the outcome. In analyzing the results (thirty weights), you determine the first and third quartiles (Q_1, Q_3) and the standard deviation (s). After completing the computations, you discover that you had incorrectly noted the lowest weight—it should have been even smaller. Will the correction change Q_1, Q_3, or s? Which ? Why?

3. Assume an intelligence test that yields IQ's that are normally distributed in the general American population, with $\bar{X} = 100$ and $s = 16$.

   a. What proportion of the population would have IQ's greater than 116?

   b. What proportion would have IQ's below 68?

   c. What proportion would have IQ's between 76 and 124?

4. Suppose that you divided the members of Boy Scout Troop 500 into two random halves (by any chance procedure you like) and then noted the individual shoe sizes of the boys in each half. On which measure of variability would you expect the two subgroups to be most alike, the interquartile range or the standard deviation? Explain.

5. A survey of all the markets in a particular city reveals that the mean price of Klimon cheese is $1.20 per pound (with $s = .24$) and the mean price of Etsaton cheese is $5.00 per pound (with $s = .50$). Which product is more variable in price, relative to its average price?

6. An instructor in automotive mechanics found that the two sections of his class differ in performance on a test of mechanical aptitude. The two sets of scores are summarized thus:

	Section A	Section B
$\bar{X}$	48	52
s	8	2

Which statistic ($\bar{X}$ or s) probably will interest him most in preparing for his instruction? Why?

7. A spelling test consisting of 100 words was administered to the class. The resulting mean score was 95 and the standard deviation was 12. What can you guess about the shape of the distribution?

8. A test on current affairs is administered to seniors in two schools. For school A, the standard deviation computed from the original scores was 6.0. For school B, the standard deviation was 6.1, computed from grouped scores with $i = 12$. What must be done before the two groups are compared for variability? Which group shows more variability in knowledge of current affairs?

DERIVED SCORES

Everywhere we turn, the performances of people, objects, and processes are scored. The instances of this phenomenon range widely—our test results, the gas mileage of our cars, the behavior of the stock market, and even our amorous adventures of the night before. In this universal behavior we perform two separate operations. First we assign a description to the individual performance, in terms of whatever unit of measurement we are using (number of items correct, inches, frequency of occurrence, and so on). This is called the *raw score*. Then we give some meaning to the raw score in terms of its adequacy, typicality, or importance.

A raw score has little, if any, meaning taken alone. It is given meaning by comparison. You would not be impressed by Johnny's report that "I got fifteen words correct on the spelling test" if you learn that the test included a hundred words and most of Johnny's classmates spelled eighty or more words correctly. Similarly, what if a friend, returning from Las Vegas, answers your "how did you make out" query with, "I won a $25 jackpot on the slot machines"? You probably will want to ask him at least another question before you congratulate him.

The comparisons that give meaning to raw scores fall into two broad categories: comparison to a specific standard of performance, and comparison to the distribution of the performances of a group. In comparisons of the first kind a particular criterion is established and goodness of the individual raw score is judged against that criterion. The following examples, quite different, should make this category clear:

1. You are trying to lose weight and your criterion is that each week for the next two months your weight must be less than that of the preceding week.

2. In order to qualify for a position you must be able successfully to perform any ten of fifteen designated tasks.

3. For admission to Alatri Academy you must have a high school grade-point average of 3.00 or more.

4. In many states, if the alcohol content of your blood reaches a specific percentage point, you are intoxicated.

5. In order to become a full-fledged member of The Illiciteers Gang, you must have committed four successful burglaries.

6. One of the authors, while serving in a World War II Parachute Infantry Regiment, failed the test for membership in an unofficial elite paratrooper society—he could not drink a quart of a creative mixture (medicinal alcohol, sake, and pineapple juice) in thirty seconds.

As these examples suggest, we can provide meaning for a score by casting it against an established criterion score. It is often possible and appropriate to do so. There are some difficulties in such comparisons, however. You may note in the illustrations that the criterion score (the standard) may be set somewhat arbitrarily. Who can say, for instance, that a grade-point average of 3.00 is any more predictive of college success than an average of 2.95? Or, would I be any less effective a gang member with only three burglaries to my credit? Nonetheless, judging a score against a specific criterion is often a necessary, if somewhat arbitrary, operation. An obvious instance of its appropriateness, despite its arbitrariness, is in setting and enforcing what is lawful behavior.

Our alternative approach to giving meaning to a raw score is to compare the individual score to a distribution of scores. This is probably the most frequent approach, perhaps because the specification of a particular criterion's performance often is not appropriate, not necessary, or not possible. There are several techniques for interpreting a score in terms of a distribution. They all involve transforming the raw score into a new score, a score derived from the raw score in such a way as to suggest the relative status of the performance in a given distribution of performances.

Of the many possible techniques, three are very frequently used: *rank order, percentile ranks,* and *standard scores.*

RANK ORDER

Probably the simplest of all methods for giving more meaning to a raw score is to indicate where the score ranks in a collection of scores. The position a score holds in the overall order is called its **rank order.** All scores are arranged in order, from highest to lowest, and each is assigned a rank order, using one for the highest, two for the next highest, and so on to the lowest score in the collection. In Table 5.1 we illustrate the procedure with the outcome of a modest poker game. The raw scores are the amounts won (or lost) by the seven players.

The table also shows how ranks are assigned to tied scores. Note that Art and Joe, with similar losses, are tied. If their raw scores (losses) were not equal, they would have been assigned ranks 5 and 6. We handle this situation by assigning

TABLE 5.1 Winnings and Ranks of Seven Players in a Game of Poker

Players	Amount won (in dollars)	Rank
Bob	11.20	1
Red	10.30	2
Bud	2.50	3
Jack	−4.80	4
Art	−5.50	5.5
Joe	−5.50	5.5
Dick	−8.20	7

each the mean of the two ranks involved (5 and 6); that is, ranks 5 and 6 are averaged to 5.5.

As we can see, Bob can boast not only that he won $11.20, but also that he was the top ranking player in the game. Red is second best—and poor Dick is ranked seventh, no doubt attributing his low ranking to bad luck rather than lack of skill.

Transforming raw scores into rank order has some advantages. As we have noted, the concept and procedure for ranking are simple and generally understood. The use of ranking also permits a rough comparison of scores by the same individual from different performances. For example, Dick might be able to say that, while it is unfortunately true that he ranked last in the poker game, he ranked second out of thirty in his tennis tournament last year.

There are, however, some disadvantages in the use of rank order. A major disadvantage is that the ranking of raw scores ignores and hides the magnitude of differences between scores. In Table 5.1, note that Bob, Red, and Bud are ranked 1, 2, and 3. However, the rankings obscure the fact that Bob and Red are relatively close in amounts won, whereas Bud is a very poor third. That is, while the differences between ranks always are equal (by definition), differences between raw scores may not be. There is also an awkwardness in using the rank-order transformation when a large number of scores is involved because, as N gets larger, the number of ties usually increases, diminishing the advantage of ranking. The reader already may have sensed also that the rank order of a score, without knowledge of the total number of scores, has relatively little meaning or at least is ambiguous.

Despite these disadvantages, expressing raw scores as ranks has wide usage. This derived score is appropriate in situations where N is small, where easy understanding is particularly important, and where the differences in magnitude between the ranked scores are not important to the purposes at hand.

PERCENTILE RANK

The **percentile rank** expresses raw scores in terms of the percent of scores that fall below it in a given group.

Let us begin by explaining the meaning of **percentile.** In formal terms, the xth percentile of a distribution of scores is the point on the scale of scores below which x percent of the scores lie. Thus, the 18th percentile is the point below which 18 percent of the scores fall and the 50th percentile is the point below which 50 percent of the scores fall (that is, the median of a distribution is the 50th percentile.) The general custom is to use the letter P to designate percentile and a numeral subscript to indicate the particular percentile. For example, P_{25} means the 25th percentile.

If this sounds somewhat familiar, your memory serves you well. In Chapter 2, we showed you how to construct a cumulative relative proportion distribution. Look back at Figure 2.5. On the vertical axis, we plotted proportions and showed that, by drawing a line from a given proportion to the curve and then reading straight down to the score scale, we could identify the score point below which that proportion of the other scores fall. If you change the vertical axis in Figure 2.5 to read percents instead of proportions, you can see that such a graphic distribution can be used to identify any percentile point in a distribution of scores.

The percentile rank (PR) for a given raw score indicates which percentile point is associated with that score. Thus, if Johnny Appleseed's raw score on a test is such that 40 percent of the scores are lower, we assign his score a percentile rank of 40; that is, PR = 40. This is a way of saying that his raw score falls approximately at P_{40} in this distribution of scores. Perhaps you can now see that Figure 2.5 in Chapter 2 also can be used to identify percentile ranks. To do so, we find the particular score (salaries, in that case) in which we are interested, read up to the curve and across to the vertical axis, locating the approximate percentage (the PR).

A graphical distribution like Figure 2.5 is convenient for estimating percentile ranks when the number of scores and cases are very large. It yields fairly good estimates. If N is small, however, you will get better approximations with the procedure illustrated in Table 5.2. The first column shows each of the thirty scores, and the real limits for each score are given in the second column. The third column shows the frequency associated with each score and, in the fourth column, the frequencies are cumulated to the upper real limit of each score. The fifth column, Cum. *fm*, is something new, however. We wish to find the percentile rank of a *score,* which you will recall, is the midpoint of its real limits. The cumulative frequency, however, is to the upper real limit of the score, not to the midpoint. Hence, we need to cumulate to the *middle* of each score interval. This is done in the Cum. *fm* column, using the familiar necessary assumption that all scores in an interval are distributed uniformly throughout the interval. Take score 37, for example, which was achieved by three persons. Theoretically, that score really can be any value from 36.5 to 37.5. We assume that the three 37's are uniformly distributed in this interval, one exactly in the middle and one exactly on either side. The procedure for computing the Cum. *fm* values follows that assumption. The cumulative frequency to the midpoint of an interval Cum. *fm* is found by adding half of the

TABLE 5.2 Computation of Percentile Ranks Corresponding to 30 Scores

Scores	Real limits	f	Cum. f	Cum. fm	PR
43	42.5–43.5	1	30	29.5	98
42	41.5–42.5	2	29	28	93
41	40.5–41.5	3	27	25.5	85
40	39.5–40.5	5	24	21.5	72
39	38.5–39.5	4	19	17	57
38	37.5–38.5	4	15	13	43
37	36.5–37.5	3	11	9.5	32
36	35.5–36.5	4	8	6	20
35	34.5–35.5	2	4	3	10
34	33.5–34.5	0	2	2	7
33	32.5–33.5	1	2	1.5	5
32	31.5–32.5	1	1	0.5	2

frequency (f) for this interval to the Cum. f value of the interval below it. Thus, the Cum. fm for the score 33 is found by taking one-half of its associated f (half of 1) and adding that value (0.5) to the cumulative frequency (Cum. f) of the interval below it (1), giving us 1.5.

Once the Cum. fm values are found, the percentile ranks are computed for each score by taking the Cum. fm as a percentage of the total, that is, by dividing each Cum. fm by 30 and rounding to the nearest whole percent. The values in the PR column are the results. We can read, for example, that the percentile rank for a score of 40 is 72. This tells us that approximately 72 percent of the thirty scores are lower than 40—which is to say that, in this distribution, $P_{72} = 40$ (approximately).

Percentile ranks and percentiles are used frequently in reporting relative performance, particularly in academic institutions. They have the advantage of being rather easily understood.

STANDARD SCORES

A more sophisticated method of converting a raw score to a relative score utilizes the standard deviation. It is relatively simple to grasp and utilize, if the meaning and computation of the standard deviation are understood. The **standard deviation** gives meaning to a raw score by comparing it to the mean of the distribution and by indicating how far the score deviates from the mean in terms of standard deviation units. If this sounds familiar, you correctly recall the discussion around formula (4.9) in Chapter 4.

For illustration, we use some interesting data supplied by an unnamable[1] researcher. He found an unusual (and happy) town in which all adult residents drank bottled beer frequently. He carefully (and, we trust, with circumspection) obtained a record of the number of bottles consumed by each adult during a thirty-day period. For each person, therefore, he had a raw score—the number of bottles of beer he or she drank in the period of time. The raw score for Steve Sknird was 200, an impressive achievement. But *how* impressive compared with the scores of other adults in this convivial town? The researcher tells us that the mean number of bottles consumed was 140. So we know that Mr. Sknird was above average. How much above average? We are told that the standard deviation for the distribution of scores was 30. Therefore, we can say that Sknird's score of 200 is two standard deviations (60 bottles) above the mean. Or we may express his score simply as +2.0, the sign indicating the direction from the mean. This kind of score sometimes is called a *sigma score* and conventionally is symbolized by the familiar lower case letter z. Thus, for our new friend Mr. Sknird, $z = +2.0$, which was computed from formula (4.9).

$$z = \frac{X - \bar{X}}{s}$$

where X is any score, $\bar{X}$ is the mean, and s is the standard deviation.

What is the z score for Polly Prub, who worked her way through only ninety-five bottles? Substituting in the formula, we have

$$z = \frac{95 - 140}{30} = \frac{-45}{30} = -1.5$$

Polly's z score is negative, indicating that her performance is below the mean, and the size of the score tells us that she is 1.5 standard deviations below the mean.

How about Ian Ylleb, whose score was 140? Since his raw score is exactly the same as the mean, can you state his score without using the formula? If not, substitute in formula (5.1) as follows:

$$z = \frac{140 - 140}{30} = \frac{0}{30} = 0.0$$

We see that Ian's z score is 0.0, indicating that it is at the mean, that is, zero distance from the mean.

Although z scores have some usage, the more frequent practice is to use a further transformation—a version that removes the necessity of the plus or minus signs and the decimal points, which are easy to forget and, in any event, awkward. Examine

---

[1] Because he doesn't exist.

Figure 5.1. The first line shows z-score points from −5.0 to +5.0. We can get rid of the decimal points by multiplying each z value by 10. Then if we add 50, we obtain a new standard score scale with a mean of 50 (instead of zero) and a standard deviation unit of 10 (instead of 1). These standard scores[2] frequently are symbolized by Z and the computational formula reads like this:

$$Z = 10 \left(\frac{X - \bar{X}}{s}\right) + 50 \qquad (5.1)$$

The second line of Figure 5.1 shows this transformation and its relationship to z scores. Why add 50 instead of any other value? Recall that we wish to eliminate negative values. Adding 50 moves the zero point from the middle of the distribution to a point five standard deviations to the left. It is very unlikely that one will encounter scores to the left of our new zero point. However, the use of 50 is otherwise quite arbitrary.

What is Polly Prub's z score? Her z score was −1.5. Multiplying by 10 and adding 50, we obtain Z = 35. Since we know that the mean Z = 50 and the standard deviation of Z = 10, we readily can see that Polly is below the mean by 1½ standard deviations.

Standard scoring with the mean as 50 and the standard deviation as 10 is popular and found often in scoring standardized tests used in elementary and secondary schools. Another transformation, also symbolized by Z, is used in Scholastic Aptitude Tests and the Graduate Record Examinations. In scoring these, the Z-score mean is set at 500 and the standard deviation of Z is 100. That is, z is multiplied by 100 and 500 is added. This transformation provides three-digit scores without decimal points. The transformation is shown on the third line of Figure 5.1. In this

z scores	-5.0	-4.0	-3.0	-2.0	-1.0	0.0	+1.0	+2.0	+3.0	+4.0	+5.0
Z score ($\bar{X} = 50$, $s = 10$)	0	10	20	30	40	50	60	70	80	90	100
Z score ($\bar{X} = 500$, $s = 100$)	0	100	200	300	400	500	600	700	800	900	1000

FIGURE 5.1   Comparisons of z-scores and two forms of Z-scores.

[2] The symbol T sometimes is used for this purpose; the literature unfortunately is inconsistent here. We prefer Z, reserving T to denote the application of this scale to normalized distributions.

system, Polly Prub's raw score of 95 would be shown as $Z = 350$, which was computed as follows:

$$Z = 100 \left(\frac{X - \bar{X}}{s} \right) + 500$$

$$= 100 \left(\frac{95 - 140}{30} \right) + 500 \qquad (5.2)$$

$$= 100 (-1.5) + 500$$

$$= -150 + 500$$

$$= 350$$

It is obvious that we have used the same symbol (Z) to denote two different standard score transformations, one with mean at 50 and one with mean at 500. In fact, there are many more variations of Z scores; you may create your own by multiplying by any constant and adding any constant. The lesson is that you should make clear *which* system you are using so that no misinterpretation results.

The Standard Scores in Comparisons and Composites

Under certain conditions, standard scores have an advantage not found in ranks or percentile ranks. Scores from different tests can be compared and even combined with standard scores, if distributions have the same shape.[3] To illustrate, assume that Frances has taken two tests, a test of verbal aptitude and a test of quantitative aptitude. The raw score means and standard deviations for the two tests are shown below, along with Frances's score (X) on each test:

	Verbal	Quantitative
$\bar{X}$	80	30
s	12	6
X	86	36

How do Frances's two scores compare? We convert the two scores to z's; thus

$$Z \text{ (verbal)} = \frac{X - \bar{X}}{s} = \frac{86 - 80}{12} = \frac{6}{12} = +0.5$$

$$Z \text{ (quantitative)} = \frac{X - \bar{X}}{s} = \frac{36 - 30}{6} = \frac{6}{6} = +1.0$$

If the two distributions are approximately similar in shape, we can say that she performed relatively better in quantitative aptitude than in verbal aptitude ($z =$

---

[3] Recall or review the discussion of distribution shapes in Chapter 2.

+ 1.0 versus $z = 0.5$). That is, standard scoring permits the comparison of apples and oranges, with similarly shaped distributions. Furthermore, under such conditions, the two scores can be added or averaged into a composite score when they are in standardized score form. Hence, we can say that Frances's combined z-score on the two tests is $0.5 + 1.0 = 1.5$, or that the mean of the two scores is 0.75, that is, $(0.5 + 1.0)/2$. Such comparisons and composites cannot be produced from ranks because N's may differ and because the score distances between ranks may not be equal. Nor can composites be derived from percentile ranks because distances between percentile ranks do not correspond to distances between associated score points.

Standard Scores and Percentile Ranks

When the distribution of raw scores is normal, there is an invariable relationship between percentile ranks and standard scores. You should already know this relationship from your recollection of the discussion in Chapter 4 on properties of the normal distribution and the explanation of Appendix II, Areas Under the Normal Curve. Any given z-score distance from the mean cuts off a given percentage of the area under a normal distribution. Verify, therefore, from Appendix II that, if the distribution is normal, a z score of $+1.0$ is equivalent to a percentile rank of 84, that is, 84 percent of the scores will be lower. Appendix II shows that 34 percent ($.3414$) of the area lies between the mean and $+1.00$; we add to this the 50 percent below the mean to get 84 percent. It is the case, however, and alas, that very few distributions of raw data are normal, so this insight has limited exact application. One often finds, on the other hand, that distributions approximate normality, in which case rough approximations of PR's can be obtained from Appendix II.

EXERCISES

1. The ten show horses in a certain exhibitor's stable are shown below with the number of blue ribbons won by each horse last year.

Horse	Number of blue ribbons	Rank	Horse	Number of blue ribbons	Rank
Valiant	12		Sally	5	
Crystal	11		Danny	3	
Uma-san	11		Cavallo	3	
Dandy Billy	11		Paisano	1	
Copper Bars	10		Wrangler	0	

(Problem continues on next page.)

    a. Give each horse a rank-order score.

    b. If someone knew only the rank-order scores, might he or she draw wrong conclusions about the relative achievements of some horses?

2. From the cumulative percent graph (Figure 5.2) answer the following questions:

    a. Is the score distribution unimodal or bimodal?

    b. What is the score point for P_{45}?

    c. What is the score point for Q_1 or P_{25}?

    d. What is the percentile rank (PR) of a score of 10?

    e. What is the PR for a score of 8.5?

    f. What percent of scores fall above a score of 14?

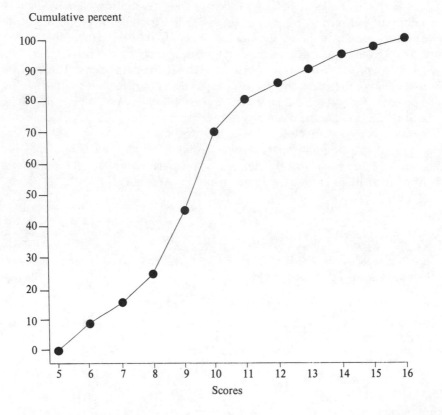

FIGURE 5.2. Cumulative Percent Graph

3. A large group of applicants for a position were given a test made up of three components: verbal ability, quantitative ability, and logical reasoning. The raw score means and standard deviations for the three subtests are given below:

	Verbal ability	Quantitative ability	Logical reasoning
$\overline{X}$	86	120	16
s	12	20	4

a. Ted Trams' scores were: verbal, 92; quantitative, 135; logical, 20. On which test did he perform relatively highest? On which was he relatively lowest? Assuming the three distributions to be comparable, what is Ted's total score?

b. Mary Niarb's score on the verbal test was 104. Express her raw score as a z score. Assuming the distribution to be normal, which is the approximate PR of her raw score?

c. The score obtained by Tony Tsep on logical reasoning was 80. Convert that score to a Z score with $\overline{X} = 50$, $s = 10$. Is this an unusual score? How do you know?

d. Sam Selims scored 122 on the test of quantitative ability. What is his Z score in the transformation with the mean as 500 and the standard deviation as 100?

DESCRIBING RELATIONSHIPS
BETWEEN VARIABLES

You have learned that the helter-skelter of scores on any variable can be simplified into statements about the shape of the distribution, central tendency, and variability. The several statistical descriptors we have presented are useful in making summary sense out of a set of measures collected on a single variable.

One other common descriptive task remains. Very often we are confronted with the question of whether the measures on one variable are related to the measures on another variable. The phenomenon is that of **correlation:** The measures on two different variables are said to be correlated when changes on one variable are associated with changes on the other. To use as an example one of the highest correlations found in nature, changes in temperature are closely associated with the number of chirps per minute from a cricket—the higher the temperature, the more frequent the chirps. Our lives are filled with assumptions about correlation between variables. Some of these assumptions square with fact (for example, intelligence test scores are correlated with academic performance to some degree). Some assumptions, alas, have little or undemonstrated adherence to fact (for example, the relationship of weight and jocularity, the relationship between race and intelligence).

CORRELATION AND CAUSATION

It is important to realize (as many people unfortunately do not) that the demonstration of correlation between two variables cannot imply, per se, that one change caused the other. As an easy example, consider a recent report that the incidence of diagnosis of lung cancer in the United States is dramatically correlated with an increase in the sale of telephones. Are you tempted to conclude that one caused the other? Or, as we hope, are you more likely to assume that the increases in both are associated with changes in other phenomena occurring at the same time? Causality is a most complex matter; it cannot be inferred merely from the existence of correlation. Return, for instance, to the very high correlation between temperature and cricket chirps. You would not conclude that thermometers are influenced by cricket chirps because you know something about the factors that produce temperature change—and cricket chirping is not consistent with that knowledge. Similarly, if you know a great deal about crickets, you might find it sensible to think that the change in temperature caused the change in the crickets' behavior. Note, however, that the correlation alone should not lead you to an inference about causation. Such an inference can be made only with additional information and with careful reasoning. Put formally, we say that the existence of correlation is a necessary but not sufficient condition for inferring cause.

THE COEFFICIENT OF CORRELATION

There are several statistical techniques for describing the degree to which two variables are associated. Each produces an index called a **coefficient of correlation** or a **correlation coefficient.** The absolute value of this coefficient may range from .00, indicating absence of correlation, through 1.00, indicating a perfect association. When the relationship between two variables is such that an increase in the value of one is associated with an *increase* in the value of the other, the correlation is said to be *positive* (or *direct*) and the coefficient carries a positive sign, which usually is omitted. When, however, the relationship is such that an increase in the value of one variable is associated with a *decrease* in the value of the other, the relationship is called *negative* (or *inverse*) and the coefficient is prefixed with a negative sign. For example, the correlation between IQ and achievement test scores tends to be about .50. This coefficient tells us that an increase in IQ is moderately associated with an increase in achievement test performance. When, however, we see a correlation coefficient of −.50, we are being told that the relationship between the two variables is such that an increase in one is moderately associated with a decrease in the other.

THE SPEARMAN RANK CORRELATION

A British psychologist, Charles Spearman, derived this correlational procedure at the beginning of this century. Because it has the virtue of simplicity, we begin with it. As its name indicates, the **Spearman rank correlation** produces a coefficient expressing the amount of correlation between two sets of rankings. To utilize this technique, therefore, we must express the obtained scores in terms of their rank order. Examine Table 6.1. You will find two sets of measures for a group of students. The second column gives the GPA for each student; the third column gives his age expressed in months. In the fourth column (A), the GPA's are ranked, using 1 for the highest GPA and proceeding through 12 for the lowest. The ages are ranked in the fifth column (B), using 1 for the highest age. In the next to last column (*D*), we calculate the difference between each pair of ranks, and, finally, we square each difference in the last column. We now have all the information needed.

TABLE 6.1. Grade-Point Averages and Ages in Months for Twelve Graduating High School Students: Computations of Rank-Order Correlation Coefficient

Student	GPA	Age (in months)	GPA rank (A)	Age rank (B)	D (A−B)	D²
A	3.4	202	4	11	−7	49
B	2.8	212	7	6	1	1
C	3.9	204	2	10	−8	64
D	2.6	224	8	1	7	49
E	3.0	217	6	3	3	9
F	2.5	210	9	7	2	4
G	2.0	215	11	4	7	49
H	1.7	198	12	12	0	0
I	3.6	207	3	9	−6	36
J	3.2	209	5	8	−3	9
K	4.0	213	1	5	−4	16
L	2.2	220	10	2	8	64
						$\Sigma D^2 = 350$

$$\rho = 1 - \frac{(6\Sigma D^2)}{N(N^2 - 1)}$$

$$= 1 - \frac{6(350)}{12(143)}$$

$$= 1 - 1.22$$

$$= -.22$$

The formula for the Spearman rank correlation coefficient, which is often symbolized by the Greek letter rho, is

$$\rho = 1 - \frac{6\Sigma D^2}{N(N^2 - 1)} \tag{6.1}$$

where D is the difference between paired rank orders on the two variables and N is the number of individuals ranked. The formula tells us to find the sum of the squared differences (ΣD^2) and to multiply this sum by 6, which gives us the numerator $(6\Sigma D^2)$. The denominator consists of the number of individuals (N) multiplied by $N^2 - 1$. The last step in computation is to divide the numerator by the denominator and subtract the result from 1. Below Table 6.1 we have substituted the appropriate numbers in formula (6.1) and find that the rank correlation between GPA and age for our twelve students is –.22. This tells us that, in this group, the relationship between grades and age is low. The minus sign tells us that the relationship is negative—that is, there is a slight tendency for older seniors to have lower GPA's.

Applicability of the Rank Coefficient

The Spearman rho (ρ) is useful and appropriate when the number of cases (N) is small—approximately thirty or less. If N is large, the Spearman procedure is awkward and open to doubt. With large N's, there is likelihood of tied ranks; if the number of tied ranks is appreciable, Spearman's rho should not be used because the technique assumes no ties. When there is a large number of tied ranks or, in any event, when N is large, other correlational techniques are more appropriate. One of these, which is explained below, is the technique from which rho is derived and of which it is a special case.

THE PEARSON PRODUCT–MOMENT COEFFICIENT OF CORRELATION

The procedure, with so impressive a name, was developed in 1896 by Karl Pearson as a refinement of a correlation technique proposed earlier by Francis Galton. It is symbolized by r and commonly is referred to simply as *the Pearson r*. The logic and algebra in the development of the Pearson r are too complex for presentation here. However, the general rationale and application of the procedure can be outlined rather simply.

The Bivariate Frequency Distribution

The earlier chapters of this book presented descriptive procedures for looking at data derived from a single variable. Those distributions and analyses are termed *univariate*. In the case of correlation, however, we are dealing with two variables at the same time. Hence, we say that the situation is *bivariate*.[1] An appreciation of the nature of correlation coefficients and their meaning can be gained from a bivariate frequency distribution, frequently referred to as a *scatter diagram*. Let's consider an overly simplified instance. Assume that, for six elementary school children, we have two measures: age to the nearest birthday and grade level. These are shown in Table 6.2.

The bivariate distribution (or scatter diagram) for these data is shown in Figure 6.1. Consider pupil F in Table 6.2. He is six years old and in the first grade. To plot both measures simultaneously, we find his grade level on the horizontal axis, his age on the vertical axis, and the intersection of the two. The lower left-hand circle in

---

[1] Although we do not include them in this text, there are procedures for correlating and otherwise analyzing more than two variables. Hence, you may hear reference to *multivariate* analyses.

TABLE 6.2 Age to Nearest Birthday and Grade Level for Six Children

Pupil	Age	Grade
A	8	3
B	11	6
C	9	4
D	7	2
E	10	5
F	6	1

Figure 6.1(a) is the bivariate location of pupil F. Each of the remaining pupils is plotted in similar fashion. Note that in this example the six plotted points all fall on a straight line which moves from lower left to upper right. The orderliness of the plots tells us that the two measures are correlated. Indeed, in this all-too-convenient example, the correlation is perfect—each increase in age is associated with a pre- cisely equivalent increase in grade level. The plots do not deviate from the straight line. The correlation is positive: an increase in one variable is associated with an increase in the other, thereby resulting in a series of plots going from lower left to upper right.

Let's move to another example in which the relationship is not perfect. In Table 6.3, we show heights and weights for 10 boys. The weights and heights are plotted in the bivariate distribution shown in Figure 6.1(b). The trend is from lower left to upper right so we know that the relationship is positive. However, we see that the association is not perfect—the plotted points do not form a nice straight line. We have drawn a straight line through the plotted points, trying to approximate a *best fit*,[2] but as you can see, while the line reflects the general configuration of the data, the plotted scores depart from our line in varying amounts. This scatter around the line demonstrates that the relationship is imperfect.

What would a scatter diagram look like in the case of *no* association between the two measures? Figure 6.1(c) is an illustration. As you can see, the plotted points scatter all over the place and they depart dramatically from a straight-line configura- tion. No matter what straight line you may draw through the plots in Figure 6.1(c), the scatter around that line is very great. Indeed, it is clear that a person's X score is unrelated to his or her Y score.

---

[2] Lest we confuse those who are more advanced than the audience for which we write, let us confess that there really are *two* best-fitting straight lines, one for the X scores and another for the Y scores. We deliberately oversimplify here in the interests of an intuitive rather than a technically correct explanation of this correlational technique.

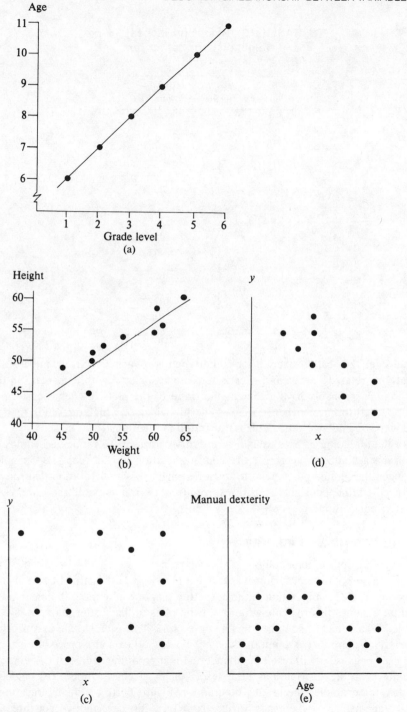

FIGURE 6.1   Illustrative scatter diagrams.

TABLE 6.3   Height (in inches) and Weight (in pounds) for Ten Boys

Boy	Height (inches)	Weight (pounds)
A	51	50
B	54	55
C	48	45
D	50	50
E	58	60
F	53	52
G	45	48
H	56	60
I	55	58
J	61	65

What does the scatter diagram look like when two variables are negatively (or inversely) correlated—when an increase in one is associated with a decrease in the other? As you probably have gathered, the trend of the plots in such a case goes from upper left to lower right. That is, an individual scoring high on X will tend to score low on Y, and vice versa. This is illustrated in Figure 6.1(d).

With the foregoing as background, an intuitive grasp of the rationale behind the Pearson r is available. You can see that a strong relationship between two variables exists when the plotted points tend to form a straight-line configuration that moves from lower left to upper right (a positive correlation) or from upper left to lower right (a negative correlation).

Assumption About Linearity

Before moving to the computation of the Pearson r, note should be taken of an important assumption underlying it. We have emphasized the straight line in our discussion of the configuration of this scatter diagram. In fact, the Pearson r procedure assumes that the configuration of the plots in the scatter diagram is best described by a straight line rather than a curved line. This is called the **assumption of linearity.** Where that assumption does not hold, the Pearson r procedure should not be used because it will result in an underestimation of the degree of correlation. Are some bivariate distributions best described by a curved line? Yes, indeed. Consider, for example, the scatter diagram in Figure 6.1(e), which plots finger dexterity against age, where age varies from birth to, say, eighty. You are not surprised to find that infants have low dexterity, that dexterity increases with age for some time and that dexterity decreases with age in later years. Hence, the best-fitting line for this relationship is curvilinear rather than linear. The plots fall closer to a curved line than to the best-fitting straight line.

There are correlational techniques for curvilinear relationships, although they are beyond the scope of this text. It is important, however, that before you compute a Pearson r you assure yourself that the assumption of linearity can fairly apply to your data. This assurance usually can be obtained by constructing a scatter diagram and inspecting the pattern that emerges. In most instances, the straight-line assumption will hold, but beware the exceptions.

Computation of the Pearson Product–Moment Coefficient

It is customary in correlational analysis to designate one variable as X and the other as Y. Bearing that in mind, we can consider a formula for the computation of the Pearson Product–Moment Coefficient of Correlation. Several alternatives, all exactly equivalent, are available. The following is our preference, assuming that a calculator is available to handle the tedium of the operations and large numbers:

$$r = \frac{\dfrac{\Sigma XY}{N} - \bar{X}\bar{Y}}{\sqrt{\left(\dfrac{\Sigma X^2}{N} - \bar{X}^2\right)\left(\dfrac{\Sigma Y^2}{N} - \bar{Y}^2\right)}} \tag{6.2}$$

where X refers to scores on one variable, Y refers to scores on the other variable, $\bar{X}$ is the mean of X scores, and $\bar{Y}$ is the mean of Y scores.

Formula (6.2) looks more formidable than it really is. If you read it carefully, you will find no surprises and nothing complex. Let's take it piece by piece. In the numerator, $\Sigma XY/N$ tells you to multiply each X score by its associated Y score, sum up their products, and divide the sum by the number of cases. From this result, subtract the product of the two means $(\bar{X}\bar{Y})$. In the denominator consider first the terms inside the first parentheses $(\Sigma X^2/N - \bar{X}^2)$, which instruct you to square each X score, sum these squares, divide by N, and then subtract the square of the mean of X. Do the same operations for Y, multiply together, find the square root of the product, and, finally, divide the numerator by the denominator.

If you are still puzzled, an example should give back your composure. Consider the data in Table 6.4, which represent the ages of 10 mothers and the ages of their oldest children. The columns and the totals give all the information required to compute with formula (6.2). Below the table, we have substituted the required numbers in the formula and have found that the correlation of mother and child age in this group is very high: $r = .99$. You can see, then, that while the computation of r entails some tedious work, it is not severe in complexity. With even a small calculator, you can produce a coefficient with reasonable dispatch once the tabular data have been organized, as in Table 6.4.

TABLE 6.4 Ages of Ten Mothers (X) and Their Oldest Children (Y): Computation of r

Mother's age (X)	Child's age (Y)	X^2	Y^2	XY
23	2	529	4	46
26	4	676	16	104
27	5	729	25	135
29	5	841	25	145
32	6	1024	36	192
35	8	1225	64	280
39	10	1521	100	390
41	11	1681	121	451
43	13	1849	169	559
44	15	1936	225	660

$\Sigma X = 339$ $\Sigma Y = 79$ $\Sigma X^2 = 12001$ $\Sigma Y^2 = 785$ $\Sigma XY = 2962$

$\overline{X} = 33.9$ $\overline{Y} = 7.9$

$$r = \frac{\dfrac{\Sigma XY}{N} - \overline{X}\,\overline{Y}}{\sqrt{\left(\dfrac{\Sigma X^2}{N} - \overline{X}^2\right)\left(\dfrac{\Sigma Y^2}{N} - \overline{Y}^2\right)}}$$

$$= \frac{2962/10 - (33.9)(7.9)}{\sqrt{\left(\dfrac{12001}{10} - (33.9)^2\right)\left(\dfrac{785}{10} - (7.9)^2\right)}}$$

$$= \frac{28.39}{28.62}$$

$$r = .99$$

Shortcuts in Computation of r

We have not presented short methods for computing the coefficient of correlation, as we did in the case of the mean and standard deviation. There are such shortcuts but, for the Pearson r, the operations remain rather cumbersome and complex. Hence, we suggest that it may be just as efficient and no less laborious to compute the Pearson coefficient directly from the original scores, using formula (6.2). A small calculator will make for less pain and fewer errors. When the scores are numerically large, however, you may and should reduce them by subtracting a constant from each. This reduction in no way affects the result. If you subtract a constant from each score on each variable and apply formula (6.2) to the new scores, the resulting coefficient will be the same as that produced by using the original scores.

Interpretation of *r:* How Large Is Large?

There is no ambiguity in interpreting $r = .00$ or $r = 1.00$. The first means no association and the second identifies perfect association. However, such results are unlikely. What can we say about the meaning of $r = .40$? What is the difference between r's of .50 and .60? Nontechnical answers to such questions are difficult to produce. One way of interpreting the meaning of r is to square it. The quantity r^2, sometimes called the **coefficient of determination,** can be read as a proportion or, if multiplied by 100, as a percent. For example, if $r = .50$, $r^2 = .25$. We can therefore say that a Pearson r of .50 can be interpreted as .25, or 25 percent. But 25 percent of *what?* The question is difficult to answer but you may get usable and adequate meaning if you think of r^2 as indicating the percent of variation in one variable that can be explained or accounted for by variation in the other. If the Pearson r between an intelligence test and success in school is .50, its square (.25) indicates that 25 percent of the variation in school success can be accounted for or explained by variation in intelligence test scores (and vice versa), leaving 75 percent (that is, $1 - r^2$) of the variation[3] in school success to be accounted for by other factors. This latter portion of the variance is said to be unexplained. Another way of putting it is that 25 percent of the total variance in each of these two measures is held in common, or *common variance*. Here, you can see that, although a correlation of .50 is halfway between $r = .00$ and $r = 1.00$, its explanatory power is much lower. A coefficient of .80 suggests that 64 percent of the variation in one variable can be attributable to variation in the other. Where two variables are related on the order of $r = .20$, the coefficient of determination ($r^2 = .04$) tells us that variation in one variable does not give us very much information about the other.

OTHER CORRELATION TECHNIQUES

We have presented two methods for arriving at an index of the relationship between two variables: the Spearman rho and the Pearson r. They will serve adequately for most bivariate situations where the assumption of linearity is reasonable and where we are interested in a description of given data rather than in inferring from our data to a larger population.

The following are examples of situations requiring different correlational procedures:

1. When the relationship between two variables tends to be curvilinear rather than linear

2. When the observations on one or both variables are qualitative rather than quantitative (referring to a distinction made in Chapter 1)

3. When we are interested in determining the correlation between a variable and the combination of two or more other variables

---

[3] This difference, reasonably enough, sometimes is referred to as the *coefficient of nondetermination*.

4. When our interest is in finding the correlation between two variables with the influence of another variable held constant.

For these, and many other special situations, there are procedures for producing coefficients of correlation. You may find them in standard statistics textbooks. Some will be understandable, from what you have learned thus far; others will require further study on your part.

EXERCISES

1. If you should examine the relationship between chronological age and reading ability, using all the pupils in an elementary school, you almost certainly would find that the correlation is very high and positive. However, if you confined your analysis to only one grade level in that same school, you would find that the correlation between age and reading is very low, and sometimes negative. Why is this the case?

2. A task of social adjustment was given to ten members of a Boy Scout troop and the Scout leader was asked, independently, to rank the ten boys in terms of his estimate of their popularity. The following are the results, expressed as ranks. What is the Spearman rank-order coefficient here?

Boy	Popularity rank	Social adjustment rank
A	1	3
B	2.5	1
C	2.5	2
D	4	8
E	5	6
F	6	4
G	7	5
H	8	7
I	9	10
J	10	9

3. It appears that, among people taking tests, those who have very high anxiety or no anxiety tend to do less well than those who are only moderately anxious. Sketch out a hypothetical scatter diagram for this situation. Why is Pearson r inappropriate for finding an index of correlation here?

4. Describe each of the following relationships as either positive or negative.

   a. Level of fatigue versus distance jogged.

b. Mental alertness versus amount of sleep deprivation.

c. Elevation above sea level versus amount of oxygen in the atmosphere.

d. Ages of married women versus ages of their spouses.

5. The existence of correlation between two variables, it was noted, is not evidence, per se, that one caused the other. Other information and reasoning must be employed to make guesses about cause. Use each of the following correlations as an exercise in inferring about cause. Which factor or factors might explain each correlation?

   a. A positive correlation between rise in washing machine sales and increase in subscriptions to the local newspaper.

   b. A positive correlation between the socioeconomic status of pupils and their scores on achievement tests.

   c. Among men, a negative correlation between length of employment in a company and the length of sickness absence.

6. We show below some data on a horse race, lifted from the pages of a newspaper. (The names of the horses have been changed to avoid embarrassment to the also-rans.) The first column identifies the entries; the second shows the order in which each horse finished the race (the winner is ranked 1); the third column tells us the betting odds for each horse before the race started (the favorite has the lowest odds). What is the correlation between betting odds and race results in this race?

Horse	Finishing order	Odds
Grande Marnier	1	5.10
Jame's Annie	2	11.40
Born Winner	3	8.60
Flicked Bick	4	2.60
Blessed Event	5	9.00
Dandy Billy	6	6.70
Pastrami Joy	7	6.00
Money Bags	8	19.40
My Tuition	9	12.80
Good Loser	10	20.70
Nobody's Friend	11	70.90
Oh My	12	63.10

7. In the following table, you see the final examination scores and scores on a quantitative aptitude test for students in an algebra course.

   a. Make a scatter diagram to see that the relationship is linear.

(Problem continues on next page.)

b. What is the coefficient of correlation for these two sets of scores?

Student	Final exam	Quantitative aptitude
1	68	41
2	84	50
3	70	40
4	110	80
5	80	49
6	98	81
7	75	46
8	103	88
9	100	68
10	86	54
11	88	58
12	102	67
13	69	43
14	73	44
15	86	56
16	100	60
17	82	46
18	90	47
19	96	75
20	84	50
21	88	57
22	93	58
23	91	47
24	105	70
25	89	50
26	78	44
27	86	48

7

INFERENCE:
GOING BEYOND DESCRIPTION

We now have completed our short tour of the field of statistical description. If, as we intended, the passage has been reasonably clear, you have learned how to summarize a given set of observations in terms of its distributional characteristics, its central tendency, and its variability. You have seen also that a particular observation can be scored relative to the variability of the total set and, finally, that the relationship between two sets of observations on the same group of individuals or objects can be summarized by an index of correlation. These elementary statistical concepts and techniques should be helpful to those who have need for summarizing data that come before them. They also should permit more critical understanding of the statistics that daily bombard us from the press, the air waves, and the speaker's platform.

However, although we promised only a treatment of elementary descriptive statistics, we are not yet done. We conclude this little book by returning to its beginning, where we left a bemused Eman A. Ton dangling sheepishly between his descriptions and his inferences. The fact is that man does not live by description alone. He generalizes and infers from what is directly known to what is not known. Everything we know or think we know about people, processes, and things is based on samples (good or poor) of people, processes, and things. The error of Eman A. Ton was not that he generalized about Ecalponians from the approximately seventy-five he met; his error was that he generalized uncritically, incautiously, and with supreme confidence.

Statistical inference is a large and complex field of theory and techniques. It is also an exciting and absorbing arena, as we hope many readers will discover by further and systematic study. We cannot do justice to it here in summary form. However, even a small taste of the task of statistical inference will alert you to the need for care and tentativeness in stretching descriptive statistics to inferences about larger groups—that is, about populations. In the following brief and closing excursion, we will discuss two aspects of statistical inference: (1) inferring about a

91

population from a formal sample of that population and (2) problems in extending statistical descriptions of nonrandom samples. Most readers of this book probably will not find themselves engaged in the first aspect. However, a little understanding of inferring from formal samples will illuminate the second aspect—which is relevant to all readers.

INFERENCES ABOUT POPULATIONS FROM SAMPLES

Let's begin with some definitions for three terms that are common in statistical inference: *population, parameter, statistic, sample.* By a **population** we mean the totality of elements (for example, objects, individuals) about which we wish to make inferences. For example: all Nomel cars manufactured in 1978, American registered voters, home owners in the City of Enalytrid, cameras imported from Japan. By a **parameter** we refer to a measure of a population. A parameter contrasts with a statistic. We have already dealt with such statistics as the mean, median, variance, and correlation coefficient. When these measures are of populations, they are called parameters; when they are about samples, they are called **statistics.** The term **sample** in statistical inference and in research work, is more difficult to define briefly. Let us start with what it is *not*. A sample is *not* a mirror image of the population from which it is drawn. It is not possible to form a sample that is exactly representative of the population. Hence, the phrase "representative sample" is erroneous, despite its frequent use. A good sample is one characterized by *impartiality* in its selection. It is selected without known bias or preference. However, this presence of impartiality will not insure representativeness, as you will see below. Impartiality here means the operation of chance rather than intentional or unconscious bias. Using some good theory about the behavior of chance events, the statistician can make some reasonable estimates of population parameters, indicating what the magnitude of error might be in those estimations. Sampling techniques achieve impartiality by the use of *randomness.*

Random Sampling

A **random sample** is one drawn in such a way that every element in the population has an equal chance of being included.[1] For an easy illustration, consider how you might draw a random sample of ten persons from a population of 100 members of your bowling club. The requirements of random sampling are closely met[2] if you put 100 slips of paper (each with a member's name) into the proverbial hat and,

---

[1] More precisely, and technically correct, a random sample of a given number of elements is drawn in such a way that all possible samples of the same size have an equal chance of being selected from the population.

[2] We say *closely met* because of a consideration beyond our purposes here. Namely, it makes a theoretical difference whether or not you replace each drawn slip in the hat before drawing the next. Our basic point is not harmed, however.

blindfolded, pick out ten slips. It is unlikely that any club member would complain if not selected.

Random selection from a small population is a simple matter. With large populations, which is usually the case, this simple random procedure is at best unwieldy and often quite impossible. A number of sampling techniques are available for use with large populations and you may learn about them through any of the many published texts on research methods in various fields. Almost all these techniques, however, use the principle of randomness at some point, thereby remaining reasonably faithful to the requirement that each element (or sample size) have an equal chance of being included. One such procedure is relatively simple and appropriate here. Confronted with the task of selecting a random sample of the 10,000 registered voters in your district, how would you proceed? Put 10,000 names in a large box and pick out your sample blindfolded? Of course not; the labor and tedium involved are punitive—and unnecessary. In such an instance a common procedure is to select one name from the voter list randomly and then include every *n*th name in the rest of the list. If, for instance, you wished a sample of 100 voters from the 10,000, select one randomly and pick every hundredth name thereafter—returning to the beginning of the list if necessary. How do you select the first name randomly? How about using today's date? Or asking a passerby to give you a number? Or using the last digit (or digits) in yesterday's New York Stock Exchange total shares traded? You can be inventive here, so long as there is no way in which personal bias or expectation about the outcome affects the choice.

The principle of random sampling, for large populations, is not easy to satisfy. Considerable care and thought must go into the effort to insure that each element has an equal chance for inclusion in the sample. The classic and frequently cited example of how things can go wrong is the *Literary Digest* poll of 1936. In that year the editors of the periodical attempted to predict the outcome of the presidential election (Roosevelt versus Landon). Their sample included approximately 10,000,000 people, which you must agree is a very large sample. On the basis of the preferences revealed by the sample, they forecast a victory for Landon. However, Roosevelt won by a landslide. What went awry? Can 10 million Americans be wrong? The error was in making selections primarily from telephone directories. Did all American opinion holders have an equal chance of being included? Obviously nontelephone subscribers were excluded from the sample. The large sample, therefore, was biased in favor of telephone subscribers, who were better off financially than nontelephone subscribers and heavily inclined toward the Republican candidate, Mr. Landon.

The cogent point about random selection is that, by providing that each population element has an equal chance of being selected, we insure that the sample is *unbiased*. Chance alone determines inclusion, not the selector's convenience, whims, prejudices, or unconscious preferences. However, while randomness insures against bias, it cannot insure representativeness in the mirror sense. The latter point is a crucial understanding. To clarify it, let's indulge in a little theoretical illustration. Let's assume that the mean age of our 100 bowling club members is 28 years and that their ages range from 19 to 35. The parameter mean, then, is 28. Suppose that we randomly select a sample of ten members and compute the mean age of that sample. Then, after returning our sample back to the total pool, we draw another random sample from the 100 members and compute the mean age of that sample. Now pretend that we repeat this process an unbelievably large number of times. We would end up with a large number of sample means (ages). Would they all be the same? Almost certainly not. In fact, we would find a distribution of sample means, some very much lower than the true (parameter) mean, some much higher, many close to the parameter mean, and some exactly the same as the parameter value. Is it possible that one of our samples will have a mean age as low as 19.5 when the population mean is actually 28? Yes, indeed—but very infrequently. That is, a sample mean far from the parameter mean is unlikely *but possible,* if the principle of randomness is permitted to operate. We would find the largest number of sample means close to the population mean, as you might guess. Indeed, if we compute the mean of the sample means, we would find that it is equal to the population mean.[3]

When we sample a population, of course, we select only one sample, not an unbelievably large number—and surely not an infinite number. Therefore, while the odds are good that our single sample statistic is closer to the parameter statistic than very far from it, we cannot be certain that this is the case. Hence, as we said, random selection of a sample, done carefully, protects against bias but cannot insure truth.

The theory and techniques of statistical inference have the task of estimating parameter values from sample statistics. These estimations take many forms. Some inference techniques tell us how much confidence we can have in estimating the parameter value as falling within a specified range of values. Others tell us how likely we are to be wrong in assuming that a parameter correlation or a parameter difference between groups is larger than zero.

EXTENDING STATISTICAL DESCRIPTIONS
OF NONRANDOM SAMPLES

We have touched, very lightly, on the process of inferring parameters of a well-defined population from the statistics derived from an unbiased sample of that population. Inferential theory and techniques are valuable and necessary knowledge for those who undertake to estimate the characteristics of defined populations; for example, political opinion pollsters, quality control specialists, market surveyors, census takers, scientific researchers. Most of us, in our daily conclusions, do not begin with defined populations which are to be sampled and whose parameters are to be estimated. Rather, in the general human condition, we are confronted with a particular set of observations, which come to us more accidentally than randomly. We describe the characteristics of this set of observations, using such

---

[3] To state theory more correctly, the mean of an *infinite* number of sample means is equal to the parameter mean. Furthermore, theoretically, the distribution of sample means is a normal distribution.

techniques as those presented earlier. Logically, we should stop at that point. That is, we should conclude that the statistical descriptions of the set apply to that particular set only and not to other possible sets. However, still in the human condition, we typically do not limit our conclusions to the given observations. We tend, rather, to go beyond description; we tend to generalize beyond what we have directly observed in our nonrandom samples—samples that were not selected to exclude bias, whether deliberate or accidental. All of us, then, go beyond statistical description. We can do so uncritically or we can be careful and cautious in our generalizations.

The discussion of the requirement of random sampling gives us one antidote for careless generalizing. We said there that an unbiased sample is one selected in such a way that every element of the population has an equal chance of being included. We can apply the lesson of that requirement in trying to decide how far we can extend our sample conclusions. Specifically, we can do some estimating of the ways in which our sample may be biased; we can try to think of the kinds of elements that had no chance of inclusion. In so doing, we are learning what restrictions must be placed on our generalizations.

For instance, let us take an overly simple exercise. You are traveling cross-country in the first-class section of an airplane and, through conversation, you find that all eighteen passengers in your section feel that welfare benefits are too high in this country. You have a firm descriptive statistic: 100 percent of the sample ($N = 18$) feels that welfare benefits are too high. Would you be tempted to estimate that all Americans feel this way? We doubt it. Nor, we hope, would you even make that estimate for all passengers on this airplane. Why? "Because," you tell us, "first-class travelers are not representative of all Americans or even of all those on the airplane—they tend to be more affluent." That is to say that less affluent people had little or no chance of being in your sample. Now, to make a new and important point, we ask whether a sample biased toward affluence would have particular views about welfare. "Yes," you say, and correctly. Therefore, you will limit your generalization severely because the sample was biased in the direction of affluence *and because affluence is related to views on welfare.*

Let us use another and less obvious illustration. Recently, a person (whose name we happily repress) conducted a survey (which fortunately will not be published). He was interested in finding out the activities that school children pursue during the summer vacation in his community. To that end, he carefully constructed a questionnaire and, with the cooperation of a theater manager, administered it to all school-aged children who attended a movie on one Saturday afternoon. He took the results, analyzed them carefully, and drew conclusions about summer activities of school children in the community. We can put his survey to our bias test. What school children did not have equal chance of being included in his sample, and is it likely that those excluded would differ in their summer activities? You can come up with more answers to that question than we note here. How about children who cannot afford the price of admission? How about those who were working? Those who were bedridden? Those who prefer outdoor activities to movies? Those who did not like the particular movie? From these questions alone, you must conclude that the surveyer was brash in his generalization. He should have limited his inferences to a much more restricted population.

But the foregoing is largely negative advice. You are asked to think of what elements had *no* chance of inclusion in your sample. Can positive advice be given? We believe so. As long as you *will* generalize beyond your sample, we suggest that you be guided by the caution that "these statistics apply for my sample and for any other group that has similar, relevant characteristics." This sounds like imprecise advice because it necessarily is imprecise. The thinking process involved has two components:

1. You eliminate from the limits of your generalization all population elements that had no chance of inclusion and that differ from your sample in characteristics that are probably related to the phenomenon you are examining.

2. You note carefully all those characteristics of your sample that are relevant to the phenomenon you are examining.

Putting the two together, you can realize some general conception of how far you conscionably can go beyond the sample statistics.

There is ample precedent for the foregoing advice. Physicians, in reporting the results of successful or unsuccessful treatments, present case studies that carefully describe the relevant characteristics of the patient. Teachers, in reporting the results of a teacher technique, carefully describe the characteristics of the sample pupils, community, instructional environment, and so on.

There is both good morality as well as intelligent behavior in extending your sample statistics only as far as common sense permits. More nonsense has been perpetuated by *overextension* than by *underextension,* and more hurt and more harm.

The conclusion? Describe carefully, generalize cautiously and reluctantly, and be prepared to change your generalization readily. If you should encounter Eman A. Ton, remind him of this with our thanks and apologies.

A REVISIT TO NUMBERS

For many people, arithmetic rules and conventions are merely painful and incomplete memories. If you are one of these unhappy souls, be assured that you have much company. It is a sad commentary on the way numbers and algebra have been taught in the educational establishment.

This Appendix reviews some elementary rules and operations you need in the text. If you already are confidently competent in the domain, we invite you to turn to Chapter 2 (after glancing at the exercises at the end of this Appendix) to confirm your self-opinion.

NOTATIONS FOR MULTIPLICATION

Somewhere along the line, you learned to express multiplication with either the times sign ($\times$) or a dot. For example, $3 \times 2 = 6$ or $3 \cdot 2 = 6$. However, in algebra, the convention is to show multiplication by juxtaposition of the terms to be multiplied. For example, to indicate that the value of X is to be multiplied by 3, we write $3X$. The expression abc tells us to multiply the values of a, b, and c. Juxtaposition alone, however, is not always sufficient. To take an absurd instance, to juxtapose 2 and 5 would produce 25 or 52. The use of parentheses removes our embarrassment here. To show that 2 is to be multiplied by 5, we write 2(5) or 5(2). To show that the quantity X is to be multiplied by 6, we write $6X$. Parentheses also permit the expression of a more complex multiplication statement: Add 2 and 5 and multiply the result by 6, can be noted unambiguously by the expression $6(2 + 5)$.[1]

Most of our needs to indicate multiplication are satisfied by the use of juxtaposition and parentheses. Occasionally, however, we use brackets to avoid the confusion of parentheses within parentheses. Consider the instruction: Add 2 and 5, multi-

---

[1] Or $(2 + 5)6$, if you insist. However, the convention is to place the least complex term first.

ply the sum by 6, and square the result. That instruction can be symbolized by $[6(2 + 5)]^2$.

NOTATIONS FOR DIVISION

The operation of division is indicated by the use of bar. Divide 6 by 2 may be expressed as $\frac{6}{2}$. But you already know this, of course. Less familiar, perhaps, is the use of the slash. For example, 6/2 also tells us to divide 6 by 2—the numerator comes before the slash, the denominator follows it. Both bar and slash are used in this book. We sometimes even use the familiar ÷ sign to show division.

OPERATIONS WITH SIGNED NUMBERS

A number may have positive or negative value. Positive numbers are indicated either by a plus sign or by the absence of a sign (for example, +5, or merely 5). Negative numbers (not surprisingly) are indicated by a minus sign, for example, –5.

When adding numbers having the same sign, we know the sum will also have the same sign. Hence, the addition of +5 and +5 produces +10. Similarly, the addition of –5 and –5 gives us –10. To put both examples in formal notation (illustrating also another use of parentheses), we have +5 + (+5) = +10 and –5 + (–5) = –10.

In adding two numbers that have unlike signs, we find the difference between the numbers and affix the sign of the larger number. For example, +5 ("I have five dollars") and – 10 ("I owe ten dollars") add up to – 5 ("I'm in trouble"). Or, 5 + (–10) = –5. When there are several numbers with unlike signs to be added, we first add those with like signs and proceed as above. Example: –2 + 5 – 3 + 4 + 6 = –5 + 15 = 10.

To subtract signed numbers, change the sign of the number to be subtracted and add. Illustration: 8 – (+4) = 8 + (–4) = 4. Similarly, 2 – (–3) = 2 + (+3) = 5.

The multiplication of numbers with similar signs produces a positive product. Hence, 5(5) = 25 and –5(–5) also is 25. When the numbers to be multiplied have unlike signs, the product is negative: 5(–5) = –25. Now see if you can discover why both of the following algebraic statements are correct:

$$2(-2)(-2)(2) = 16$$

but

$$2(-2)(-2)(-2) = -16$$

The same principles are followed in the division of numbers with similar or unlike signs. Thus

$$32/8 = 4$$
$$-49/-7 = 7$$
$$-21/3 = -7$$
$$18/-9 = -2$$

WORKING WITH FRACTIONS

To add or subtract fractions that have the same denominator, add or subtract the numerators and retain the same (common) denominator. Illustrations: $1/3 + 1/3 = 2/3$ and (using symbols) $5/b - 2/b = 3/b$. How about $1/4 - (1-1/4)?$[2]

To add or subtract fractions that have different denominators, we first must make the denominators alike (find a *common denominator*). To do this, we recall that the value of a fraction is not changed if we multiply both the numerator and the denominator by the same number. If, for example, we multiply both the numerator and the denominator of $1/2$ by 3, we get $(1/2)(3/3) = 3/6 = 1/2$. To add $1/2 + 1/4$, we note that if the first fraction is multiplied by $2/2$, it can be made "common" with the second. Therefore, we have $(1/2)(2/2) + 1/4 = 2/4 + 1/4 = 3/4$.

In the case of $1/3 + 2/7$, however, we must find a new denominator common to both. There are many possibilities, but it is most convenient to find the lowest one. In this instance, the product of the two denominators (21) is the lowest. Thus, $(1/3)(7/7) + (2/7)(3/3) = 7/21 + 6/21 = 13/21$. You can always determine a common denominator simply by multiplying the denominators together in this manner. However, this does not always yield the lowest value. Consider, for example, $1/6 + 4/9$. The product of the two denominators is 54. Is this the smallest number into which both 6 and 9 can be divided evenly? Both 18 and 36, also, can be divided evenly by 6 and 9. Since we want the *lowest* value, we will use 18 as our common denominator. In order to convert both fractions to the denominator 18, we multiply the first fraction by $3/3$ and the second by $2/2$. The operations look like this:

$$1/6 + 4/9 = (1/6)(3/3) + (4/9)(2/2)$$

$$= 3/18 + 8/18 = 11/18$$

Multiplication of fractions depends upon a simple rule: Multiply the numerators and multiply the denominators. Thus, $(1/3)(2/5) = 2/15$, and (to use unlike signs) $(1/3)(-2/5) = -2/15$.

In dividing one fraction by another, invert the fraction in the denominator and proceed as in the case of multiplication:

$$\frac{1/3}{2/5} = 1/3(5/2) = 5/6$$

How do you multiply or divide a fraction by a whole number? Remember that any whole number can be expressed as a fraction merely by placing it over 1. In other words, $5 = 5/1$. Thus

$$(3/8)(2) = (3/8)(2/1) = 6/8$$

$$(3/8)/2 = (3/8)/(2/1) = (3/8)(1/2) = 3/16$$

---

[2] Let's pause to avoid confusing you. We wrote the last fraction as $-1/4$. Do we mean $\frac{-1}{4}$ or $-\frac{1}{4}$? If you think for a moment, you will see that it makes no difference. In either case we have a negative value; dividing -1 by 4 yields a negative result.

Fractions frequently are *reduced* or *simplified* by dividing both the numerator and the denominator by the same value. Again, the overall value of the fraction is unaffected by this procedure. Taking 3/6 as an example, both 3 and 6 are divisible by 3, giving us $\dfrac{3 \div 3}{6 \div 3} = 1/2$. Similarly,

$$14/35 = (14/7)/(35/7) = 2/5$$

$$15/18 = (15/3)/(18/3) = 5/6$$

Dividing by the largest possible value results in the greatest possible reduction in the fraction. Thus, although both the numerator and the denominator of 60/90 are divisible by 2 (and, indeed, by many other numbers), dividing by 30 reduces the fraction to 2/3, which is as far as it can be reduced.[3]

WORKING WITH DECIMALS

When multiplying numbers involving decimals, first multiply the numbers as though there were no decimals, then count the total number of decimal places (to right of decimal points) and mark off that number of decimal places in the product: 0.5(0.3) = 0.15 and 1.3(0.22) = 0.286.

To divide a decimal by a decimal, first turn the denominator (the number you divide by) into a whole number by moving the decimal point to the end of that number; then move the decimal point in the numerator an equivalent number of places to the right. Using this rule, 2.4/1.2 = 24/12 = 2. Similarly, 0.984/0.08 = 98.4/8 = 12.3.

ROUNDING

When rounding a number, we must examine the value beyond the digit to which we are rounding. If the value beyond this digit is greater than 5, increase this digit by 1; if it is less than 5, leave this digit as is. For example,

$$0.846 \text{ to the nearest hundredth} = 0.85$$

$$1.28 \text{ to the nearest tenth} = 1.3$$

$$4.8 \text{ to the nearest one} = 5$$

$$28 \text{ to the nearest ten} = 30$$

No doubt you now are wondering what to do when the value beyond this digit equals 5. It is important to determine whether this value is *exactly* 5—if it is the least bit greater (or less) than 5, we proceed as discussed above. For example,

$$.250001 \text{ to the nearest tenth} = .3$$

$$.249999 \text{ to the nearest tenth} = .2$$

---

[3] Note that if we had divided both the numerator and the denominator by 2, the resulting fraction (30/45) could be reduced still further.

However, when the value beyond the digit to which we are rounding is *exactly* 5, we make use of a popular convention: round the digit to the nearest even number. Thus

$$5.17500 \text{ rounded to the nearest hundredth} = 5.18$$

$$3.45000 \text{ rounded to the nearest tenth} = 3.4$$

$$6.500 \text{ rounded to the nearest one} = 6$$

$$55 \text{ rounded to the nearest ten} = 60$$

This convention is practical in that it allows chance to decide whether a number is rounded up or down. In the long run, the rounding will occur in one direction as often as in the other.

SQUARES AND SQUARE ROOTS

Some descriptive statistics require the use of squares and square roots. The square of a number is that number multiplied by itself (for example, $4^2 = 16$). The square root of a number, indicated by the symbol $\sqrt{}$, is the quantity that, when multiplied by itself, equals the number (for example, $\sqrt{4} = \pm 2$).[4]

Appendix III provides squares and square roots that are accurate enough for this text and the general computations of descriptive statistics. In the first column (labeled N), you will find numbers between 1.00 and 10.00. The second column (N^2) gives the squares of these numbers. The square root for each N is given in the third column ($\sqrt{N}$). The last column ($\sqrt{10N}$) permits you to read directly the square roots of numbers from 10 through 100. For example, the square root of 80 can be found by locating the 8.00 in the first column and looking across to the $10N$. That is, $10(8) = 80$. The approximate square root of 80 is found to be 8.94427. (Can you discover another way of estimating the square root of 80 from Appendix III, using the N^2 column?)

However, what about square roots for numbers smaller than 1.00 or larger than 100.00? Many electronic calculators, even small ones, will provide the answers at the touch of a fingertip. If you have one available, you can escape the following procedure—indeed, you can forget Appendix III entirely!

The approximate square roots of numbers smaller than 1.00 may be found as follows. We illustrate by determining the square root of 0.0055542.

1. Round the number to three significant digits.[5] This gives us 0.00555—if you remember that zeros following the decimal are not digits; they are place holders.

2. Move the decimal point an *even* number of places to the *right* in order to obtain a number that falls between 1.00 and 100.00 (that is, within the limits of the N and $10N$ columns of Appendix III). In the case of 0.00555, we can move the decimal point four places, obtaining 55.5.

---

[4] As you should recall, a negative value multiplied by a negative value results in a positive product. Hence, any number actually has two square roots, one positive and one negative.

[5] The selection of three significant digits is arbitrary. It results in an adequate approximation.

3. Look up the square root of 55.5 in Appendix III (you will have to use the $\sqrt{10N}$ column, that is, find $\sqrt{10(5.55)}$). The result is 7.44983.

4. The final step is to move the decimal point back to where it belongs. This is done by moving the decimal to the *left, half* as many places as we moved it originally. We first moved four places to the right; now we move two places to the left, turning 7.44983 into .0744983. Rounding the result to three significant digits gives us 0.0745 as the approximate square root of the original number, 0.0055542.

To find the square root of a number larger than 100, we use the same basic procedure, but this time we reverse direction when moving the decimal point. For example, let's find the approximate square root of 54,321.

1. Round the number to three significant digits. This gives us 54300.

2. Move the decimal point to the *left* an *even* number of places, giving us a number that falls between 1.00 and 100.00. If we move the decimal four places, we get an appropriate number: 5.43.

3. From Appendix III, we find the square root of 5.43 to be 2.33024.

4. Move the decimal point to the *right,* half as many places as we did to the left earlier. Our example requires us to move two places to the right, giving us 233.02, which is the approximate square root of 54,321, to three significant digits. (To show you that the approximation is not bad, we tell you that the correct square root, to two decimal places, is 233.07.)

The reminders and definitions presented above should suffice for confronting the presentations in this book. For those who wish it, more detailed explanation of these and other elementary operations used in statistics are available elsewhere.[6]

EXERCISES

You should be able to perform the following operations without pain. Correct answers may be found in Appendix IV.

1. $2(4 - 5) = $ _____

2. $\dfrac{X(10 + 5)}{5} = $ _____

---

[6] See Andrew R. Baggaley, *Mathematics for Introductory Statistics* (New York: John Wiley & Sons, 1969) or Helen M. Walker, *Mathematics Essential for Elementary Statistics* (New York: Holt, Rinehart & Winston, 1951). Such a review, with a highly interesting supplement to the usual treatment of statistics, is found in James K. Brewer's *Everything You Always Wanted to Know About Statistics But Didn't Know How to Ask*. (Dubuque, Iowa: Kendall-Hunt, 1978).

3. $[X(2 + 6 - 7)] [X(-8 + 4 + 5)] = $ _____

4. $\sqrt{4}(\sqrt{16}) = $ _____

5. $-4(-2) - 8(-2) = $ _____   **6.** $\dfrac{-10}{2} = $ _____

7. $\dfrac{-18}{-3} = $ _____   **8.** $\left(\dfrac{-10}{5}\right)\left(\dfrac{X}{2}\right) = $ _____

9. $\dfrac{1}{2} \div \dfrac{1}{2} = $ _____

10. $\dfrac{X(5-6)}{(4-6)} \div \dfrac{-2}{-X} = $ _____

11. $\dfrac{1}{2} + \dfrac{1}{6} = $ _____   **12.** $.12(1.2) = $ _____

13. $2.25/.5 = $ _____   **14.** $\dfrac{.8(.2)}{-.04} = $ _____

15. $\sqrt{9.66} = $ _____   **16.** $\sqrt{90} = $ _____

17. $\sqrt{.0945} = $ _____   **18.** $\sqrt{1234} = $ _____

19. $-\sqrt{64} \div -2 = $ _____

20. Round the following values:

.5616 to the nearest tenth = _____

7.8500 to the nearest tenth = _____

.7513 to the nearest tenth = _____

44.5002 to the nearest whole number = _____

.4939 to the nearest whole number = _____

25.0 to the nearest ten = _____

145 to the nearest hundred = _____

21. Express each of the following statements in mathematical form. (Example: "Add 2 and 2" is expressed as $2 + 2$.)

a. Multiply the square root of X by 3 = _____

b. Add X and Y and divide the sum by one-half = _____

c. Divide X by 5 and multiply the quotient by 2 times .05 = _____

AREAS UNDER THE NORMAL CURVE

z	Area between mean and z	Area beyond z	z	Area between mean and z	Area beyond z
1	2	3	1	2	3
0.00	.0000	.5000	0.35	.1368	.3632
0.01	.0040	.4960	0.36	.1406	.3594
0.02	.0080	.4920	0.37	.1443	.3557
0.03	.0120	.4880	0.38	.1480	.3520
0.04	.0160	.4840	0.39	.1517	.3483
0.05	.0199	.4801	0.40	.1554	.3446
0.06	.0239	.4761	0.41	.1591	.3409
0.07	.0279	.4721	0.42	.1628	.3372
0.08	.0319	.4681	0.43	.1664	.3336
0.09	.0359	.4641	0.44	.1700	.3300
0.10	.0398	.4602	0.45	.1736	.3264
0.11	.0438	.4562	0.46	.1772	.3228
0.12	.0478	.4522	0.47	.1808	.3192
0.13	.0517	.4483	0.48	.1844	.3156
0.14	.0557	.4443	0.49	.1879	.3121
0.15	.0596	.4404	0.50	.1915	.3085
0.16	.0636	.4364	0.51	.1950	.3050
0.17	.0675	.4325	0.52	.1985	.3015
0.18	.0714	.4286	0.53	.2019	.2981
0.19	.0753	.4247	0.54	.2054	.2946
0.20	.0793	.4207	0.55	.2088	.2912
0.21	.0832	.4168	0.56	.2123	.2877
0.22	.0871	.4129	0.57	.2157	.2843
0.23	.0910	.4090	0.58	.2190	.2810
0.24	.0948	.4052	0.59	.2224	.2776
0.25	.0987	.4013	0.60	.2257	.2743
0.26	.1026	.3974	0.61	.2291	.2709
0.27	.1064	.3936	0.62	.2324	.2676
0.28	.1103	.3897	0.63	.2357	.2643
0.29	.1141	.3859	0.64	.2389	.2611
0.30	.1179	.3821	0.65	.2422	.2578
0.31	.1217	.3783	0.66	.2454	.2546
0.32	.1255	.3745	0.67	.2486	.2514
0.33	.1293	.3707	0.68	.2517	.2483
0.34	.1331	.3669	0.69	.2549	.2451

From Clarke, Robert B., Coladarci, Arthur P., and Caffrey, John, *Statistical Reasoning and Procedures*. (Columbus, Ohio: Merrill, 1965). Reprinted by permission of the Charles E. Merrill Publishing Company.

z 1	Area between mean and z 2	Area beyond z 3	z 1	Area between mean and z 2	Area beyond z 3
0.70	.2580	.2420	1.05	.3531	.1469
0.71	.2611	.2389	1.06	.3554	.1446
0.72	.2642	.2358	1.07	.3577	.1423
0.73	.2673	.2327	1.08	.3599	.1401
0.74	.2704	.2296	1.09	.3621	.1379
0.75	.2734	.2266	1.10	.3643	.1357
0.76	.2764	.2236	1.11	.3665	.1335
0.77	.2794	.2206	1.12	.3686	.1314
0.78	.2823	.2177	1.13	.3708	.1292
0.79	.2852	.2148	1.14	.3729	.1271
0.80	.2881	.2119	1.15	.3749	.1251
0.81	.2910	.2090	1.16	.3770	.1230
0.82	.2939	.2061	1.17	.3790	.1210
0.83	.2967	.2033	1.18	.3810	.1190
0.84	.2995	.2005	1.19	.3830	.1170
0.85	.3023	.1977	1.20	.3849	.1151
0.86	.3051	.1949	1.21	.3869	.1131
0.87	.3078	.1922	1.22	.3888	.1112
0.88	.3106	.1894	1.23	.3907	.1093
0.89	.3133	.1867	1.24	.3925	.1075
0.90	.3159	.1841	1.25	.3944	.1056
0.91	.3186	.1814	1.26	.3962	.1038
0.92	.3212	.1788	1.27	.3980	.1020
0.93	.3238	.1762	1.28	.3997	.1003
0.94	.3264	.1736	1.29	.4015	.0985
0.95	.3289	.1711	1.30	.4032	.0968
0.96	.3315	.1685	1.31	.4049	.0951
0.97	.3340	.1660	1.32	.4066	.0934
0.08	.3365	.1635	1.33	.4082	.0918
0.99	.3389	.1611	1.34	.4099	.0901
1.00	.3413	.1587	1.35	.4115	.0885
1.01	.3438	.1562	1.36	.4131	.0869
1.02	.3461	.1539	1.37	.4147	.0853
1.03	.3485	.1515	1.38	.4162	.0838
1.04	.3508	.1492	1.39	.4177	.0823

z 1	Area between mean and z 2	Area beyond z 3	z 1	Area between mean and z 2	Area beyond z 3
1.40	.4192	.0808	1.75	.4599	.0401
1.41	.4207	.0793	1.76	.4608	.0392
1.42	.4222	.0778	1.77	.4616	.0384
1.43	.4236	.0764	1.78	.4625	.0375
1.44	.4251	.0749	1.79	.4633	.0367
1.45	.4265	.0735	1.80	.4641	.0359
1.46	.4279	.0721	1.81	.4649	.0351
1.47	.4292	.0708	1.82	.4656	.0344
1.48	.4306	.0694	1.83	.4664	.0336
1.49	.4319	.0681	1.84	.4671	.0329
1.50	.4332	.0668	1.85	.4678	.0322
1.51	.4345	.0655	1.86	.4686	.0314
1.52	.4357	.0643	1.87	.4693	.0307
1.53	.4370	.0630	1.88	.4699	.0301
1.54	.4382	.0618	1.89	.4706	.0294
1.55	.4394	.0606	1.90	.4713	.0287
1.56	.4406	.0594	1.91	.4719	.0281
1.57	.4418	.0582	1.92	.4726	.0274
1.58	.4429	.0571	1.93	.4732	.0268
1.59	.4441	.0559	1.94	.4738	.0262
1.60	.4452	.0548	1.95	.4744	.0256
1.61	.4463	.0537	1.96	.4750	.0250
1.62	.4474	.0526	1.97	.4756	.0244
1.63	.4484	.0516	1.98	.4761	.0239
1.64	.4495	.0505	1.99	.4767	.0233
1.65	.4505	.0495	2.00	.4772	.0228
1.66	.4515	.0485	2.01	.4778	.0222
1.67	.4525	.0475	2.02	.4783	.0217
1.68	.4535	.0465	2.03	.4788	.0212
1.69	.4545	.0455	2.04	.4793	.0207
1.70	.4554	.0446	2.05	.4798	.0202
1.71	.4564	.0436	2.06	.4803	.0197
1.72	.4573	.0427	2.07	.4808	.0192
1.73	.4582	.0418	2.08	.4812	.0188
1.74	.4591	.0409	2.09	.4817	.0183

z 1	Area between mean and z 2	Area beyond z 3	z 1	Area between mean and z 2	Area beyond z 3
2.10	.4821	.0179	2.45	.4929	.0071
2.11	.4826	.0174	2.46	.4931	.0069
2.12	.4830	.0170	2.47	.4932	.0068
2.13	.4834	.0166	2.48	.4934	.0066
2.14	.4838	.0162	2.49	.4936	.0064
2.15	.4842	.0158	2.50	.4938	.0062
2.16	.4846	.0154	2.51	.4940	.0060
2.17	.4850	.0150	2.52	.4941	.0059
2.18	.4854	.0146	2.53	.4943	.0057
2.19	.4857	.0143	2.54	.4945	.0055
2.20	.4861	.0139	2.55	.4946	.0054
2.21	.4864	.0136	2.56	.4948	.0052
2.22	.4868	.0132	2.57	.4949	.0051
2.23	.4871	.0129	2.58	.4951	.0049
2.24	.4875	.0125	2.59	.4952	.0048
2.25	.4878	.0122	2.60	.4953	.0047
2.26	.4881	.0119	2.61	.4955	.0045
2.27	.4884	.0116	2.62	.4956	.0044
2.28	.4887	.0113	2.63	.4957	.0043
2.29	.4890	.0110	2.64	.4959	.0041
2.30	.4893	.0107	2.65	.4960	.0040
2.31	.4896	.0104	2.66	.4961	.0039
2.32	.4898	.0102	2.67	.4962	.0038
2.33	.4901	.0099	2.68	.4963	.0037
2.34	.4904	.0096	2.69	.4964	.0036
2.35	.4906	.0094	2.70	.4965	.0035
2.36	.4909	.0091	2.71	.4966	.0034
2.37	.4911	.0089	2.72	.4967	.0033
2.38	.4913	.0087	2.73	.4968	.0032
2.39	.4916	.0084	2.74	.4969	.0031
2.40	.4918	.0082	2.75	.4970	.0030
2.41	.4920	.0080	2.76	.4971	.0029
2.42	.4922	.0078	2.77	.4972	.0028
2.43	.4925	.0075	2.78	.4973	.0027
2.44	.4927	.0073	2.79	.4974	.0026

z 1	Area between mean and z 2	Area beyond z 3	z 1	Area between mean and z 2	Area beyond z 3
2.80	.4974	.0026	3.15	.4992	.0008
2.81	.4975	.0025	3.16	.4992	.0008
2.82	.4976	.0024	3.17	.4992	.0008
2.83	.4977	.0023	3.18	.4993	.0007
2.84	.4977	.0023	3.19	.4993	.0007
2.85	.4978	.0022	3.20	.4993	.0007
2.86	.4979	.0021	3.21	.4993	.0007
2.87	.4979	.0021	3.22	.4994	.0006
2.88	.4980	.0020	3.23	.4994	.0006
2.89	.4981	.0019	3.24	.4994	.0006
2.90	.4981	.0019	3.30	.4995	.0005
2.91	.4982	.0018	3.40	.4997	.0003
2.92	.4982	.0018	3.50	.4998	.0002
2.93	.4983	.0017	3.60	.4998	.0002
2.94	.4984	.0016	3.70	.4999	.0001
2.95	.4984	.0016			
2.96	.4985	.0015			
2.97	.4985	.0015			
2.98	.4986	.0014			
2.99	.4986	.0014			
3.00	.4987	.0013			
3.01	.4987	.0013			
3.02	.4987	.0013			
3.03	.4988	.0012			
3.04	.4988	.0012			
3.05	.4989	.0011			
3.06	.4989	.0011			
3.07	.4989	.0011			
3.08	.4990	.0010			
3.09	.4990	.0010			
3.10	.4990	.0010			
3.11	.4991	.0009			
3.12	.4991	.0009			
3.13	.4991	.0009			
3.14	.4992	.0008			

SQUARES AND SQUARE ROOTS

N	N^2	$\sqrt{N}$	$\sqrt{10N}$	N	N^2	$\sqrt{N}$	$\sqrt{10N}$
1.00	1.0000	1.00000	3.16228	1.50	2.2500	1.22474	3.87298
1.01	1.0201	1.00499	3.17805	1.51	2.2801	1.22882	3.88587
1.02	1.0404	1.00995	3.19374	1.52	2.3104	1.23288	3.89872
1.03	1.0609	1.01489	3.20936	1.53	2.3409	1.23693	3.91152
1.04	1.0816	1.01980	3.22490	1.54	2.3716	1.24097	3.92428
1.05	1.1025	1.02470	3.24037	1.55	2.4025	1.24499	3.93700
1.06	1.1236	1.02956	3.25576	1.56	2.4336	1.24900	3.94968
1.07	1.1449	1.03441	3.27109	1.57	2.4649	1.25300	3.96232
1.08	1.1664	1.03923	3.28634	1.58	2.4964	1.25698	3.97492
1.09	1.1881	1.04403	3.30151	1.59	2.5281	1.26095	3.98748
1.10	1.2100	1.04881	3.31662	1.60	2.5600	1.26491	4.00000
1.11	1.2321	1.05357	3.33167	1.61	2.5921	1.26886	4.01248
1.12	1.2544	1.05830	3.34664	1.62	2.6244	1.27279	4.02492
1.13	1.2769	1.06301	3.36155	1.63	2.6569	1.27671	4.03733
1.14	1.2996	1.06771	3.37639	1.64	2.6896	1.28062	4.04969
1.15	1.3225	1.07238	3.39116	1.65	2.7225	1.28452	4.06202
1.16	1.3456	1.07703	3.40588	1.66	2.7556	1.28841	4.07431
1.17	1.3689	1.08167	3.42053	1.67	2.7889	1.29228	4.08656
1.18	1.3924	1.08628	3.43511	1.68	2.8224	1.29615	4.09878
1.19	1.4161	1.09087	3.44964	1.69	2.8561	1.30000	4.11096
1.20	1.4400	1.09545	3.46410	1.70	2.8900	1.30384	4.12311
1.21	1.4641	1.10000	3.47851	1.71	2.9241	1.30767	4.13521
1.22	1.4884	1.10454	3.49285	1.72	2.9584	1.31149	4.14729
1.23	1.5129	1.10905	3.50714	1.73	2.9929	1.31529	4.15933
1.24	1.5376	1.11355	3.52136	1.74	3.0276	1.31909	4.17133
1.25	1.5625	1.11803	3.53553	1.75	3.0625	1.32288	4.18330
1.26	1.5876	1.12250	3.54965	1.76	3.0976	1.32665	4.19524
1.27	1.6129	1.12694	3.56371	1.77	3.1329	1.33041	4.20714
1.28	1.6384	1.13137	3.57771	1.78	3.1684	1.33417	4.21900
1.29	1.6641	1.13578	3.59166	1.79	3.2041	1.33791	4.23084
1.30	1.6900	1.14018	3.60555	1.80	3.2400	1.34164	4.24264
1.31	1.7161	1.14455	3.61939	1.81	3.2761	1.34536	4.25441
1.32	1.7424	1.14891	3.63318	1.82	3.3124	1.34907	4.26615
1.33	1.7689	1.15326	3.64692	1.83	3.3489	1.35277	4.27785
1.34	1.7956	1.15758	3.66060	1.84	3.3856	1.35647	4.28952
1.35	1.8225	1.16190	3.67423	1.85	3.4225	1.36015	4.30116
1.36	1.8496	1.16619	3.68782	1.86	3.4596	1.36382	4.31277
1.37	1.8769	1.17047	3.70135	1.87	3.4969	1.36748	4.32435
1.38	1.9044	1.17473	3.71484	1.88	3.5344	1.37113	4.33590
1.39	1.9321	1.17898	3.72827	1.89	3.5721	1.37477	4.34741
1.40	1.9600	1.18322	3.74166	1.90	3.6100	1.37840	4.35890
1.41	1.9881	1.18743	3.75500	1.91	3.6481	1.38203	4.37035
1.42	2.0164	1.19164	3.76829	1.92	3.6864	1.38564	4.38178
1.43	2.0449	1.19583	3.78153	1.93	3.7249	1.38924	4.39318
1.44	2.0736	1.20000	3.79473	1.94	3.7636	1.39284	4.40454
1.45	2.1025	1.20416	3.80789	1.95	3.8025	1.39642	4.41588
1.46	2.1316	1.20830	3.82099	1.96	3.8416	1.40000	4.42719
1.47	2.1609	1.21244	3.83406	1.97	3.8809	1.40357	4.43847
1.48	2.1904	1.21655	3.84708	1.98	3.9204	1.40712	4.44972
1.49	2.2201	1.22066	3.86005	1.99	3.9601	1.41067	4.46094

N	N^2	$\sqrt{N}$	$\sqrt{10N}$	N	N^2	$\sqrt{N}$	$\sqrt{10N}$
2.00	4.0000	1.41421	4.47214	2.50	6.2500	1.58114	5.00000
2.01	4.0401	1.41774	4.48330	2.51	6.3001	1.58430	5.00999
2.02	4.0804	1.42127	4.49444	2.52	6.3504	1.58745	5.01996
2.03	4.1209	1.42478	4.50555	2.53	6.4009	1.59060	5.02991
2.04	4.1616	1.42829	4.51664	2.54	6.4516	1.59374	5.03984
2.05	4.2025	1.43178	4.52769	2.55	6.5025	1.59687	5.04975
2.06	4.2436	1.43527	4.53872	2.56	6.5536	1.60000	5.05964
2.07	4.2849	1.43875	4.54973	2.57	6.6049	1.60312	5.06952
2.08	4.3264	1.44222	4.56070	2.58	6.6564	1.60624	5.07937
2.09	4.3681	1.44568	4.57165	2.59	6.7081	1.60935	5.08920
2.10	4.4100	1.44914	4.58258	2.60	6.7600	1.61245	5.09902
2.11	4.4521	1.45258	4.59347	2.61	6.8121	1.61555	5.10882
2.12	4.4944	1.45602	4.60435	2.62	6.8644	1.61864	5.11859
2.13	4.5369	1.45945	4.61519	2.63	6.9169	1.62173	5.12835
2.14	4.5796	1.46287	4.62601	2.64	6.9696	1.62481	5.13809
2.15	4.6225	1.46629	4.63681	2.65	7.0225	1.62788	5.14782
2.16	4.6656	1.46969	4.64758	2.66	7.0756	1.63095	5.15752
2.17	4.7089	1.47309	4.65833	2.67	7.1289	1.63401	5.16720
2.18	4.7524	1.47648	4.66905	2.68	7.1824	1.63707	5.17687
2.19	4.7961	1.47986	4.67974	2.69	7.2361	1.64012	5.18652
2.20	4.8400	1.48324	4.69042	2.70	7.2900	1.64317	5.19615
2.21	4.8841	1.48661	4.70106	2.71	7.3441	1.64621	5.20577
2.22	4.9284	1.48997	4.71169	2.72	7.3984	1.64924	5.21536
2.23	4.9729	1.49332	4.72229	2.73	7.4529	1.65227	5.22494
2.24	5.0176	1.49666	4.73286	2.74	7.5076	1.65529	5.23450
2.25	5.0625	1.50000	4.74342	2.75	7.5625	1.65831	5.24404
2.26	5.1076	1.50333	4.75395	2.76	7.6176	1.66132	5.25357
2.27	5.1529	1.50665	4.76445	2.77	7.6729	1.66433	5.26308
2.28	5.1984	1.50997	4.77493	2.78	7.7284	1.66733	5.27257
2.29	5.2441	1.51327	4.78539	2.79	7.7841	1.67033	5.28205
2.30	5.2900	1.51658	4.79583	2.80	7.8400	1.67332	5.29150
2.31	5.3361	1.51987	4.80625	2.81	7.8961	1.67631	5.30094
2.32	5.3824	1.52315	4.81664	2.82	7.9524	1.67929	5.31037
2.33	5.4289	1.52643	4.82701	2.83	8.0089	1.68226	5.31977
2.34	5.4756	1.52971	4.83735	2.84	8.0656	1.68523	5.32917
2.35	5.5225	1.53297	4.84768	2.85	8.1225	1.68819	5.33854
2.36	5.5696	1.53623	4.85798	2.86	8.1796	1.69115	5.34790
2.37	5.6169	1.53948	4.86826	2.87	8.2369	1.69411	5.35724
2.38	5.6644	1.54272	4.87852	2.88	8.2944	1.69706	5.36656
2.39	5.7121	1.54596	4.88876	2.89	8.3521	1.70000	5.37587
2.40	5.7600	1.54919	4.89898	2.90	8.4100	1.70294	5.38516
2.41	5.8081	1.55242	4.90918	2.91	8.4681	1.70587	5.39444
2.42	5.8564	1.55563	4.91935	2.92	8.5264	1.70880	5.40370
2.43	5.9049	1.55885	4.92950	2.93	8.5849	1.71172	5.41295
2.44	5.9536	1.56205	4.93964	2.94	8.6436	1.71464	5.42218
2.45	6.0025	1.56525	4.94975	2.95	8.7025	1.71756	5.43139
2.46	6.0516	1.56844	4.95984	2.96	8.7616	1.72047	5.44059
2.47	6.1009	1.57162	4.96991	2.97	8.8209	1.72337	5.44977
2.48	6.1504	1.57480	4.97996	2.98	8.8804	1.72627	5.45894
2.49	6.2001	1.57797	4.98999	2.99	8.9401	1.72916	5.46809

N	N^2	$\sqrt{N}$	$\sqrt{10N}$	N	N^2	$\sqrt{N}$	$\sqrt{10N}$
3.00	9.0000	1.73205	5.47723	3.50	12.2500	1.87083	5.91608
3.01	9.0601	1.73494	5.48635	3.51	12.3201	1.87350	5.92453
3.02	9.1204	1.73781	5.49545	3.52	12.3904	1.87617	5.93296
3.03	9.1809	1.74069	5.50454	3.53	12.4609	1.87883	5.94138
3.04	9.2416	1.74356	5.51362	3.54	12.5316	1.88149	5.94979
3.05	9.3025	1.74642	5.52268	3.55	12.6025	1.88414	5.95819
3.06	9.3636	1.74929	5.53173	3.56	12.6736	1.88680	5.96657
3.07	9.4249	1.75214	5.54076	3.57	12.7449	1.88944	5.97495
3.08	9.4864	1.75499	5.54977	3.58	12.8164	1.89209	5.98331
3.09	9.5481	1.75784	5.55878	3.59	12.8881	1.89473	5.99166
3.10	9.6100	1.76068	5.56776	3.60	12.9600	1.89737	6.00000
3.11	9.6721	1.76352	5.57674	3.61	13.0321	1.90000	6.00833
3.12	9.7344	1.76635	5.58570	3.62	13.1044	1.90263	6.01664
3.13	9.7969	1.76918	5.59464	3.63	13.1769	1.90526	6.02495
3.14	9.8596	1.77200	5.60357	3.64	13.2496	1.90788	6.03324
3.15	9.9225	1.77482	5.61249	3.65	13.3225	1.91050	6.04152
3.16	9.9856	1.77764	5.62139	3.66	13.3956	1.91311	6.04979
3.17	10.0489	1.78045	5.63028	3.67	13.4689	1.91572	6.05805
3.18	10.1124	1.78326	5.63915	3.68	13.5424	1.91833	6.06630
3.19	10.1761	1.78606	5.64801	3.69	13.6161	1.92094	6.07454
3.20	10.2400	1.78885	5.65685	3.70	13.6900	1.92354	6.08276
3.21	10.3041	1.79165	5.66569	3.71	13.7641	1.92614	6.09098
3.22	10.3684	1.79444	5.67450	3.72	13.8384	1.92873	6.09918
3.23	10.4329	1.79722	5.68331	3.73	13.9129	1.93132	6.10737
3.24	10.4976	1.80000	5.69210	3.74	13.9876	1.93391	6.11555
3.25	10.5625	1.80278	5.70088	3.75	14.0625	1.93649	6.12372
3.26	10.6276	1.80555	5.70964	3.76	14.1376	1.93907	6.13188
3.27	10.6929	1.80831	5.71839	3.77	14.2129	1.94165	6.14003
3.28	10.7584	1.81108	5.72713	3.78	14.2884	1.94422	6.14817
3.29	10.8241	1.81384	5.73585	3.79	14.3641	1.94679	6.15630
3.30	10.8900	1.81659	5.74456	3.80	14.4400	1.94936	6.16441
3.31	10.9561	1.81934	5.75326	3.81	14.5161	1.95192	6.17252
3.32	11.0224	1.82209	5.76194	3.82	14.5924	1.95448	6.18061
3.33	11.0889	1.82483	5.77062	3.83	14.6689	1.95704	6.18870
3.34	11.1556	1.82757	5.77927	3.84	14.7456	1.95959	6.19677
3.35	11.2225	1.83030	5.78792	3.85	14.8225	1.96214	6.20484
3.36	11.2896	1.83303	5.79655	3.86	14.8996	1.96469	6.21289
3.37	11.3569	1.83576	5.80517	3.87	14.9769	1.96723	6.22093
3.38	11.4244	1.83848	5.81378	3.88	15.0544	1.96977	6.22896
3.39	11.4921	1.84120	5.82237	3.89	15.1321	1.97231	6.23699
3.40	11.5600	1.84391	5.83095	3.90	15.2100	1.97484	6.24500
3.41	11.6281	1.84662	5.83952	3.91	15.2881	1.97737	6.25300
3.42	11.6964	1.84932	5.84808	3.92	15.3664	1.97990	6.26099
3.43	11.7649	1 85203	5 85662	3.93	15.4449	1.98242	6.26897
3.44	11.8336	1.85472	5.86515	3.94	15.5236	1.98494	6.27694
3.45	11.9025	1.85742	5.87367	3.95	15.6025	1.98746	6.28490
3.46	11.9716	1.86011	5.88218	3.96	15.6816	1.98997	6.29285
3.47	12.0409	1.86279	5.89067	3.97	15.7609	1.99249	6.30079
3.48	12.1104	1.86548	5.89915	3.98	15.8408	1.99499	6.30872
3.49	12.1801	1.86815	5.90762	3.99	15.9201	1.99750	6.31664

N	N^2	$\sqrt{N}$	$\sqrt{10N}$	N	N^2	$\sqrt{N}$	$\sqrt{10N}$
4.00	16.0000	2.00000	6.32456	4.50	20.2500	2.12132	6.70820
4.01	16.0801	2.00250	6.33246	4.51	20.3401	2.12368	6.71565
4.02	16.1604	2.00499	6.34035	4.52	20.4304	2.12603	6.72309
4.03	16.2409	2.00749	6.34823	4.53	20.5209	2.12838	6.73053
4.04	16.3216	2.00998	6.35610	4.54	20.6116	2.13073	6.73795
4.05	16.4025	2.01246	6.36396	4.55	20.7025	2.13307	6.74537
4.06	16.4836	2.01494	6.37181	4.56	20.7936	2.13542	6.75278
4.07	16.5649	2.01742	6.37966	4.57	20.8849	2.13776	6.76018
4.08	16.6464	2.01990	6.38749	4.58	20.9764	2.14009	6.76757
4.09	16.7281	2.02237	6.39531	4.59	21.0681	2.14243	6.77495
4.10	16.8100	2.02485	6.40312	4.60	21.1600	2.14476	6.78233
4.11	16.8921	2.02731	6.41093	4.61	21.2521	2.14709	6.78970
4.12	16.9744	2.02978	6.41872	4.62	21.3444	2.14942	6.79706
4.13	17.0569	2.03224	6.42651	4.63	21.4369	2.15174	6.80441
4.14	17.1396	2.03470	6.43428	4.64	21.5296	2.15407	6.81175
4.15	17.2225	2.03715	6.44205	4.65	21.6225	2.15639	6.81909
4.16	17.3056	2.03961	6.44981	4.66	21.7156	2.15870	6.82642
4.17	17.3889	2.04206	6.45755	4.67	21.8089	2.16102	6.83374
4.18	17.4724	2.04450	6.46529	4.68	21.9024	2.16333	6.84105
4.19	17.5561	2.04695	6.47302	4.69	21.9961	2.16564	6.84836
4.20	17.6400	2.04939	6.48074	4.70	22.0900	2.16795	6.85565
4.21	17.7241	2.05183	6.48845	4.71	22.1841	2.17025	6.86294
4.22	17.8084	2.05426	6.49615	4.72	22.2784	2.17256	6.87023
4.23	17.8929	2.05670	6.50384	4.73	22.3729	2.17486	6.87750
4.24	17.9776	2.05913	6.51153	4.74	22.4676	2.17715	6.88477
4.25	18.0625	2.06155	6.51920	4.75	22.5625	2.17945	6.89202
4.26	18.1476	2.06398	6.52687	4.76	22.6576	2.18174	6.89928
4.27	18.2329	2.06640	6.53452	4.77	22.7529	2.18403	6.90652
4.28	18.3184	2.06882	6.54217	4.78	22.8484	2.18632	6.91375
4.29	18.4041	2.07123	6.54981	4.79	22.9441	2.18861	6.92098
4.30	18.4900	2.07364	6.55744	4.80	23.0400	2.19089	6.92820
4.31	18.5761	2.07605	6.56506	4.81	23.1361	2.19317	6.93542
4.32	18.6624	2.07846	6.57267	4.82	23.2324	2.19545	6.94262
4.33	18.7489	2.08087	6.58027	4.83	23.3289	2.19773	6.94982
4.34	18.8356	2.08327	6.58787	4.84	23.4256	2.20000	6.95701
4.35	18.9225	2.08567	6.59545	4.85	23.5225	2.20227	6.96419
4.36	19.0096	2.08806	6.60303	4.86	23.6196	2.20454	6.97137
4.37	19.0969	2.09045	6.61060	4.87	23.7169	2.20681	6.97854
4.38	19.1844	2.09284	6.61816	4.88	23.8144	2.20907	6.98570
4.39	19.2721	2.09523	6.62571	4.89	23.9121	2.21133	6.99285
4.40	19.3600	2.09762	6.63325	4.90	24.0100	2.21359	7.00000
4.41	19.4481	2.10000	6.64078	4.91	24.1081	2.21585	7.00714
4.42	19.5364	2.10238	6.64831	4.92	24.2064	2.21811	7.01427
4.43	19.6249	2.10476	6.65582	4.93	24.3049	2.22036	7.02140
4.44	19.7136	2.10713	6.66333	4.94	24.4036	2.22261	7.02851
4.45	19.8025	2.10950	6.67083	4.95	24.5025	2.22486	7.03562
4.46	19.8916	2.11187	6.67832	4.96	24.6016	2.22711	7.04273
4.47	19.9809	2.11424	6.68581	4.97	24.7009	2.22935	7.04982
4.48	20.0704	2.11660	6.69328	4.98	24.8004	2.23159	7.05691
4.49	20.1601	2.11896	6.70075	4.99	24.9001	2.23383	7.06399

N	N^2	$\sqrt{N}$	$\sqrt{10N}$	N	N^2	$\sqrt{N}$	$\sqrt{10N}$
5.00	25.0000	2.23607	7.07107	5.50	30.2500	2.34521	7.41620
5.01	25.1001	2.23830	7.07814	5.51	30.3601	2.34734	7.42294
5.02	25.2004	2.24054	7.08520	5.52	30.4704	2.34947	7.42967
5.03	25.3009	2.24277	7.09225	5.53	30.5809	2.35160	7.43640
5.04	25.4016	2.24499	7.09930	5.54	30.6916	2.35372	7.44312
5.05	25.5025	2.24722	7.10634	5.55	30.8025	2.35584	7.44983
5.06	25.6036	2.24944	7.11337	5.56	30.9136	2.35797	7.45654
5.07	25.7049	2.25167	7.12039	5.57	31.0249	2.36008	7.46324
5.08	25.8064	2.25389	7.12741	5.58	31.1364	2.36220	7.46994
5.09	25.9081	2.25610	7.13442	5.59	31.2481	2.36432	7.47663
5.10	26.0100	2.25832	7.14143	5.60	31.3600	2.36643	7.48331
5.11	26.1121	2.26053	7.14843	5.61	31.4721	2.36854	7.48999
5.12	26.2144	2.26274	7.15542	5.62	31.5844	2.37065	7.49667
5.13	26.3169	2.26495	7.16240	5.63	31.6969	2.37276	7.50333
5.14	26.4196	2.26716	7.16938	5.64	31.8096	2.37487	7.50999
5.15	26.5225	2.26936	7.17635	5.65	31.9225	2.37697	7.51665
5.16	26.6256	2.27156	7.18331	5.66	32.0356	2.37908	7.52330
5.17	26.7289	2.27376	7.19027	5.67	32.1489	2.38118	7.52994
5.18	26.8324	2.27596	7.19722	5.68	32.2624	2.38328	7.53658
5.19	26.9361	2.27816	7.20417	5.69	32.3761	2.38537	7.54321
5.20	27.0400	2.28035	7.21110	5.70	32.4900	2.38747	7.54983
5.21	27.1441	2.28254	7.21803	5.71	32.6041	2.38956	7.55645
5.22	27.2484	2.28473	7.22496	5.72	32.7184	2.39165	7.56307
5.23	27.3529	2.28692	7.23187	5.73	32.8329	2.39374	7.56968
5.24	27.4576	2.28910	7.23878	5.74	32.9476	2.39583	7.57628
5.25	27.5625	2.29129	7.24569	5.75	33.0625	2.39792	7.58288
5.26	27.6676	2.29347	7.25259	5.76	33.1776	2.40000	7.58947
5.27	27.7729	2.29565	7.25948	5.77	33.2929	2.40208	7.59605
5.28	27.8784	2.29783	7.26636	5.78	33.4084	2.40416	7.60263
5.29	27.9841	2.30000	7.27324	5.79	33.5241	2.40624	7.60920
5.30	28.0900	2.30217	7.28011	5.80	33.6400	2.40832	7.61577
5.31	28.1961	2.30434	7.28697	5.81	33.7561	2.41039	7.62234
5.32	28.3024	2.30651	7.29383	5.82	33.8724	2.41247	7.62889
5.33	28.4089	2.30868	7.30068	5.83	33.9889	2.41454	7.63544
5.34	28.5156	2.31084	7.30753	5.84	34.1056	2.41661	7.64199
5.35	28.6225	2.31301	7.31437	5.85	34.2225	2.41868	7.64853
5.36	28.7296	2.31517	7.32120	5.86	34.3396	2.42074	7.65506
5.37	28.8369	2.31733	7.32803	5.87	34.4569	2.42281	7.66159
5.38	28.9444	2.31948	7.33485	5.88	34.5744	2.42487	7.66812
5.39	29.0521	2.32164	7.34166	5.89	34.6921	2.42693	7.67463
5.40	29.1600	2.32379	7.34847	5.90	34.8100	2.42899	7.68115
5.41	29.2681	2.32594	7.35527	5.91	34.9281	2.43105	7.68765
5.42	29.3764	2.32809	7.36205	5.92	35.0464	2.43311	7.69415
5.43	29.4849	2.33024	7.36885	5.93	35.1649	2.43516	7.70065
5.44	29.5936	2.33238	7.37564	5.94	35.2836	2.43721	7.70714
5.45	29.7025	2.33452	7.38241	5.95	35.4025	2.43926	7.71362
5.46	29.8116	2.33666	7.38918	5.96	35.5216	2.44131	7.72010
5.47	29.9209	2.33880	7.39594	5.97	35.6409	2.44336	7.72658
5.48	30.0304	2.34094	7.40270	5.98	35.7604	2.44540	7.73305
5.49	30.1401	2.34307	7.40945	5.99	35.8801	2.44745	7.73951

N	N^2	$\sqrt{N}$	$\sqrt{10N}$	N	N^2	$\sqrt{N}$	$\sqrt{10N}$
6.00	36.0000	2.44949	7.74597	6.50	42.2500	2.54951	8.06226
6.01	36.1201	2.45153	7.75242	6.51	42.3801	2.55147	8.06846
6.02	36.2404	2.45357	7.75887	6.52	42.5104	2.55343	8.07465
6.03	36.3609	2.45561	7.76531	6.53	42.6409	2.55539	8.08084
6.04	36.4816	2.45764	7.77174	6.54	42.7716	2.55734	8.08703
6.05	36.6025	2.45967	7.77817	6.55	42.9025	2.55930	8.09321
6.06	36.7236	2.46171	7.78460	6.56	43.0336	2.56125	8.09938
6.07	36.8449	2.46374	7.79102	6.57	43.1649	2.56320	8.10555
6.08	36.9664	2.46577	7.79744	6.58	43.2964	2.56515	8.11172
6.09	37.0881	2.46779	7.80385	6.59	43.4281	2.56710	8.11788
6.10	37.2100	2.46982	7.81025	6.60	43.5600	2.56905	8.12404
6.11	37.3321	2.47184	7.81665	6.61	43.6921	2.57099	8.13019
6.12	37.4544	2.47386	7.82304	6.62	43.8244	2.57294	8.13634
6.13	37.5769	2.47588	7.82943	6.63	43.9569	2.57488	8.14248
6.14	37.6996	2.47790	7.83582	6.64	44.0896	2.57682	8.14862
6.15	37.8225	2.47992	7.84219	6.65	44.2225	2.57876	8.15475
6.16	37.9456	2.48193	7.84857	6.66	44.3556	2.58070	8.16088
6.17	38.0689	2.48395	7.85493	6.67	44.4889	2.58263	8.16701
6.18	38.1924	2.48596	7.86130	6.68	44.6224	2.58457	8.17313
6.19	38.3161	2.48797	7.86766	6.69	44.7561	2.58650	8.17924
6.20	38.4400	2.48998	7.87401	6.70	44.8900	2.58844	8.18535
6.21	38.5641	2.49199	7.88036	6.71	45.0241	2.59037	8.19146
6.22	38.6884	2.49399	7.88670	6.72	45.1584	2.59230	8.19756
6.23	38.8129	2.49600	7.89303	6.73	45.2929	2.59422	8.20366
6.24	38.9376	2.49800	7.89937	6.74	45.4276	2.59615	8.20975
6.25	39.0625	2.50000	7.90569	6.75	45.5625	2.59808	8.21584
6.26	39.1876	2.50200	7.91202	6.76	45.6976	2.60000	8.22192
6.27	39.3129	2.50400	7.91833	6.77	45.8329	2.60192	8.22800
6.28	39.4384	2.50599	7.92465	6.78	45.9684	2.60384	8.23408
6.29	39.5641	2.50799	7.93095	6.79	46.1041	2.60576	8.24015
6.30	39.6900	2.50998	7.93725	6.80	46.2400	2.60768	8.24621
6.31	39.8161	2.51197	7.94355	6.81	46.3761	2.60960	8.25227
6.32	39.9424	2.51396	7.94984	6.82	46.5124	2.61151	8.25833
6.33	40.0689	2.51595	7.95613	6.83	46.6489	2.61343	8.26438
6.34	40.1956	2.51794	7.96241	6.84	46.7856	2.61534	8.27043
6.35	40.3225	2.51992	7.96869	6.85	46.9225	2.61725	8.27647
6.36	40.4496	2.52190	7.97496	6.86	47.0596	2.61916	8.28251
6.37	40.5769	2.52389	7.98123	6.87	47.1969	2.62107	8.28855
6.38	40.7044	2.52587	7.98749	6.88	47.3344	2.62298	8.29458
6.39	40.8321	2.52784	7.99375	6.89	47.4721	2.62488	8.30060
6.40	40.9600	2.52982	8.00000	6.90	47.6100	2.62679	8.30662
6.41	41.0881	2.53180	8.00625	6.91	47.7481	2.62869	8.31264
6.42	41.2164	2.53377	8.01249	6.92	47.8864	2.63059	8.31865
6.43	41.3449	2.53574	8.01873	6.93	48.0249	2.63249	8.32466
6.44	41.4736	2.53772	8.02496	6.94	48.1636	2.63439	8.33067
6.45	41.6025	2.53969	8.03119	6.95	48.3025	2.63629	8.33667
6.46	41.7316	2.54165	8.03741	6.96	48.4416	2.63818	8.34266
6.47	41.8609	2.54362	8.04363	6.97	48.5809	2.64008	8.34865
6.48	41.9904	2.54558	8.04984	6.98	48.7204	2.64197	8.35464
6.49	42.1201	2.54755	8.05605	6.99	48.8601	2.64386	8.36062

N	N²	√N	√10N	N	N²	√N	√10N
7.00	49.0000	2.64575	8.36660	7.50	56.2500	2.73861	8.66025
7.01	49.1401	2.64764	8.37257	7.51	56.4001	2.74044	8.66603
7.02	49.2804	2.64953	8.37854	7.52	56.5504	2.74226	8.67179
7.03	49.4209	2.65141	8.38451	7.53	56.7009	2.74408	8.67756
7.04	49.5616	2.65330	8.39047	7.54	56.8516	2.74591	8.68332
7.05	49.7025	2.65518	8.39643	7.55	57.0025	2.74773	8.68907
7.06	49.8436	2.65707	8.40238	7.56	57.1536	2.74955	8.69483
7.07	49.9849	2.65895	8.40833	7.57	57.3049	2.75136	8.70057
7.08	50.1264	2.66083	8.41427	7.58	57.4564	2.75318	8.70632
7.09	50.2681	2.66271	8.42021	7.59	57.6081	2.75500	8.71206
7.10	50.4100	2.66458	8.42615	7.60	57.7600	2.75681	8.71780
7.11	50.5521	2.66646	8.43208	7.61	57.9121	2.75862	8.72353
7.12	50.6944	2.66833	8.43801	7.62	58.0644	2.76043	8.72926
7.13	50.8369	2.67021	8.44393	7.63	58.2169	2.76225	8.73499
7.14	50.9796	2.67208	8.44985	7.64	58.3696	2.76405	8.74071
7.15	51.1225	2.67395	8.45577	7.65	58.5225	2.76586	8.74643
7.16	51.2656	2.67582	8.46168	7.66	58.6756	2.76767	8.75214
7.17	51.4089	2.67769	8.46759	7.67	58.8289	2.76948	8.75785
7.18	51.5524	2.67955	8.47349	7.68	58.9824	2.77128	8.76356
7.19	51.6961	2.68142	8.47939	7.69	59.1361	2.77308	8.76926
7.20	51.8400	2.68328	8.48528	7.70	59.2900	2.77489	8.77496
7.21	51.9841	2.68514	8.49117	7.71	59.4441	2.77669	8.78066
7.22	52.1284	2.68701	8.49706	7.72	59.5984	2.77849	8.78635
7.23	52.2729	2.68887	8.50294	7.73	59.7529	2.78029	8.79204
7.24	52.4176	2.69072	8.50882	7.74	59.9076	2.78209	8.79773
7.25	52.5625	2.69258	8.51469	7.75	60.0625	2.78388	8.80341
7.26	52.7076	2.69444	8.52056	7.76	60.2176	2.78568	8.80909
7.27	52.8529	2.69629	8.52643	7.77	60.3729	2.78747	8.81476
7.28	52.9984	2.69815	8.53229	7.78	60.5284	2.78927	8.82043
7.29	53.1441	2.70000	8.53815	7.79	60.6841	2.79106	8.82610
7.30	53.2900	2.70185	8.54400	7.80	60.8400	2.79285	8.83176
7.31	53.4361	2.70370	8.54985	7.81	60.9961	2.79464	8.83742
7.32	53.5824	2.70555	8.55570	7.82	61.1524	2.79643	8.84308
7.33	53.7289	2.70740	8.56154	7.83	61.3089	2.79821	8.84873
7.34	53.8756	2.70924	8.56738	7.84	61.4656	2.80000	8.85438
7.35	54.0225	2.71109	8.57321	7.85	61.6225	2.80179	8.86002
7.36	54.1696	2.71293	3.57904	7.86	61.7796	2.80357	8.86566
7.37	54.3169	2.71477	8.58487	7.87	61.9369	2.80535	8.87130
7.38	54.4644	2.71662	8.59069	7.88	62.0944	2.80713	8.87694
7.39	54.6121	2.71846	8.59651	7.89	62.2521	2.80891	8.88257
7.40	54.7600	2.72029	8.60233	7.90	62.4100	2.81069	8.88819
7.41	54.9081	2.72213	8.60814	7.91	62.5681	2.81247	8.89382
7.42	55.0564	2.72397	8.61394	7.92	62.7264	2.81425	8.89944
7 43	55.2049	2.72580	8.61974	7.93	62.8849	2.81603	8.90505
7.44	55.3536	2.72764	8.62554	7.94	63.0436	2.81780	8.91067
7.45	55.5025	2.72947	8.63134	7.95	63.2025	2.81957	8.91628
7.46	55.6516	2.73130	8.63713	7.96	63.3616	2.82135	8.92188
7.47	55.8009	2.73313	8.64292	7.97	63.5209	2.82312	8.92749
7.48	55.9504	2.73496	8.64870	7.98	63.6804	2.82489	8.93308
7.49	56.1001	2.73679	8.65448	7.99	63.8401	2.82666	8.93868

N	N²	√N	√10N	N	N²	√N	√10N
8.00	64.0000	2.82843	8.94427	8.50	72.2500	2.91548	9.21954
8.01	64.1601	2.83019	8.94986	8.51	72.4201	2.91719	9.22497
8.02	64.3204	2.83196	8.95545	8.52	72.5904	2.91890	9.23038
8.03	64.4809	2.83373	8.96103	8.53	72.7609	2.92062	9.23580
8.04	64.6416	2.83549	8.96660	8.54	72.9316	2.92233	9.24121
8.05	64.8025	2.83725	8.97218	8.55	73.1025	2.92404	9.24662
8.06	64.9636	2.83901	8.97775	8.56	73.2736	2.92575	9.25203
8.07	65.1249	2.84077	8.98332	8.57	73.4449	2.92746	9.25743
8.08	65.2864	2.84253	8.98888	8.58	73.6164	2.92916	9.26283
8.09	65.4481	2.84429	8.99444	8.59	73.7881	2.93087	9.26823
8.10	65.6100	2.84605	9.00000	8.60	73.9600	2.93258	9.27362
8.11	65.7721	2.84781	9.00555	8.61	74.1321	2.93428	9.27901
8.12	65.9344	2.84956	9.01110	8.62	74.3044	2.93598	9.28440
8.13	66.0969	2.85132	9.01665	8.63	74.4769	2.93769	9.28978
8.14	66.2596	2.85307	9.02219	8.64	74.6496	2.93939	9.29516
8.15	66.4225	2.85482	9.02774	8.65	74.8225	2.94109	9.30054
8.16	66.5856	2.85657	9.03327	8.66	74.9956	2.94279	9.30591
8.17	66.7489	2.85832	9.03881	8.67	75.1689	2.94449	9.31128
8.18	66.9124	2.86007	9.04434	8.68	75.3424	2.94618	9.31665
8.19	67.0761	2.86182	9.04986	8.69	75.5161	2.94788	9.32202
8.20	67.2400	2.86356	9.05539	8.70	75.6900	2.94958	9.32738
8.21	67.4041	2.86531	9.06091	8.71	75.8641	2.95127	9.33274
8.22	67.5684	2.86705	9.06642	8.72	76.0384	2.95296	9.33809
8.23	67.7329	2.86880	9.07193	8.73	76.2129	2.95466	9.34345
8.24	67.8976	2.87054	9.07744	8.74	76.3876	2.95635	9.34880
8.25	68.0625	2.87228	9.08295	8.75	76.5625	2.95804	9.35414
8.26	68.2276	2.87402	9.08845	8.76	76.7376	2.95973	9.35949
8.27	68.3929	2.87576	9.09395	8.77	76.9129	2.96142	9.36483
8.28	68.5584	2.87750	9.09945	8.78	77.0884	2.96311	9.37017
8.29	68.7241	2.87924	9.10494	8.79	77.2641	2.96479	9.37550
8.30	68.8900	2.88097	9.11043	8.80	77.4400	2.96648	9.38083
8.31	69.0561	2.88271	9.11592	8.81	77.6161	2.96816	9.38616
8.32	69.2224	2.88444	9.12140	8.82	77.7924	2.96985	9.39149
8.33	69.3889	2.88617	9.12688	8.83	77.9689	2.97153	9.39681
8.34	69.5556	2.88791	9.13236	8.84	78.1456	2.97321	9.40213
8.35	69.7225	2.88964	9.13783	8.85	78.3225	2.97489	9.40744
8.36	69.8896	2.89137	9.14330	8.86	78.4996	2.97658	9.41276
8.37	70.0569	2.89310	9.14877	8.87	78.6769	2.97825	9.41807
8.38	70.2244	2.89482	9.15423	8.88	78.8544	2.97993	9.42338
8.39	70.3921	2.89655	9.15969	8.89	79.0321	2.98161	9.42868
8.40	70.5600	2.89828	9.16515	8.90	79.2100	2.98329	9.43398
8.41	70.7281	2.90000	9.17061	8.91	79.3881	2.98496	9.43928
8.42	70.8964	2.90172	9.17606	8.92	79.5664	2.98664	9.44458
8.43	71.0649	2.90345	9.18150	8.93	79.7449	2.98831	9.44987
8.44	71.2336	2.90517	9.18695	8.94	79.9236	2.98998	9.45516
8.45	71.4025	2.90689	9.19239	8.95	80.1025	2.99166	9.46044
8.46	71.5716	2.90861	9.19783	8.96	80.2816	2.99333	9.46573
8.47	71.7409	2.91033	9.20326	8.97	80.4609	2.99500	9.47101
8.48	71.9104	2.91204	9.20869	8.98	80.6404	2.99666	9.47629
8.49	72.0801	2.91376	9.21412	8.99	80.8201	2.99833	9.48156

N	N²	√N	√10N	N	N²	√N	√10N
9.00	81.0000	3.00000	9.48683	9.50	90.2500	3.08221	9.74679
9.01	81.1801	3.00167	9.49210	9.51	90.4401	3.08383	9.75192
9.02	81.3604	3.00333	9.49737	9.52	90.6304	3.08545	9.75705
9.03	81.5409	3.00500	9.50263	9.53	90.8209	3.08707	9.76217
9.04	81.7216	3.00666	9.50789	9.54	91.0116	3.08869	9.76729
9.05	81.9025	3.00832	9.51315	9.55	91.2025	3.09031	9.77241
9.06	82.0836	3.00998	9.51840	9.56	91.3936	3.09192	9.77753
9.07	82.2649	3.01164	9.52365	9.57	91.5849	3.09354	9.78264
9.08	82.4464	3.01330	9.52890	9.58	91.7764	3.09516	9.78775
9.09	82.6281	3.01496	9.53415	9.59	91.9681	3.09677	9.79285
9.10	82.8100	3.01662	9.53939	9.60	92.1600	3.09839	9.79796
9.11	82.9921	3.01828	9.54463	9.61	92.3521	3.10000	9.80306
9.12	83.1744	3.01993	9.54987	9.62	92.5444	3.10161	9.80816
9.13	83.3569	3.02159	9.55510	9.63	92.7369	3.10322	9.81326
9.14	83.5396	3.02324	9.56033	9.64	92.9296	3.10483	9.81835
9.15	83.7225	3.02490	9.56556	9.65	93.1225	3.10644	9.82344
9.16	83.9056	3.02655	9.57079	9.66	93.3156	3.10805	9.82853
9.17	84.0889	3.02820	9.57601	9.67	93.5089	3.10966	9.83362
9.18	84.2724	3.02985	9.58123	9.68	93.7024	3.11127	9.83870
9.19	84.4561	3.03150	9.58645	9.69	93.8961	3.11288	9.84378
9.20	84.6400	3.03315	9.59166	9.70	94.0900	3.11448	9.84886
9.21	84.8241	3.03480	9.59687	9.71	94.2841	3.11609	9.85393
9.22	85.0084	3.03645	9.60208	9.72	94.4784	3.11769	9.85901
9.23	85.1929	3.03809	9.60729	9.73	94.6729	3.11929	9.86408
9.24	85.3776	3.03974	9.61249	9.74	94.8676	3.12090	9.86914
9.25	85.5625	3.04138	9.61769	9.75	95.0625	3.12250	9.87421
9.26	85.7476	3.04302	9.62289	9.76	95.2576	3.12410	9.87927
9.27	85.9329	3.04467	9.62808	9.77	95.4529	3.12570	9.88433
9.28	86.1184	3.04631	9.63328	9.78	95.6484	3.12730	9.88939
9.29	86.3041	3.04795	9.63846	9.79	95.8441	3.12890	9.89444
9.30	86.4900	3.04959	9.64365	9.80	96.0400	3.13050	9.89949
9.31	86.6761	3.05123	9.64883	9.81	96.2361	3.13209	9.90454
9.32	86.8624	3.05287	9.65401	9.82	96.4324	3.13369	9.90959
9.33	87.0489	3.05450	9.65919	9.83	96.6289	3.13528	9.91464
9.34	87.2356	3.05614	9.66437	9.84	96.8256	3.13688	9.91968
9.35	87.4225	3.05778	9.66954	9.85	97.0225	3.13847	9.92472
9.36	87.6096	3.05941	9.67471	9.86	97.2196	3.14006	9.92975
9.37	87.7969	3.06105	9.67988	9.87	97.4169	3.14166	9.93479
9.38	87.9844	3.06268	9.68504	9.88	97.6144	3.14325	9.93982
9.39	88.1721	3.06431	9.69020	9.89	97.8121	3.14484	9.94485
9.40	88.3600	3.06594	9.69536	9.90	98.0100	3.14643	9.94987
9.41	88.5481	3.06757	9.70052	9.91	98.2081	3.14802	9.95490
9.42	88.7364	3.06920	9.70567	9.92	98.4064	3.14960	9.95992
9.43	88.9249	3.07083	9.71082	9.93	98.6049	3.15119	9.96494
9.44	89.1136	3.07246	9.71597	9.94	98.8036	3.15278	9.96995
9.45	89.3025	3.07409	9.72111	9.95	99.0025	3.15436	9.97497
9.46	89.4916	3.07571	9.72625	9.96	99.2016	3.15595	9.97998
9.47	89.6809	3.07734	9.73139	9.97	99.4009	3.15753	9.98499
9.48	89.8704	3.07896	9.73653	9.98	99.6004	3.15911	9.98999
9.49	90.0601	3.08058	9.74166	9.99	99.8001	3.16070	9.99500

ANSWERS TO EXERCISES

Chapter 2 (pages 25–28)

1.   **a.** quantitative and continuous   **b.** qualitative

      c. qualitative   **d.** quantitative and continuous

      e. quantitative and continuous   **f.** quantitative and discrete

      g. quantitative and continuous   **h.** qualitative

      i. qualitative

2.   **a.**

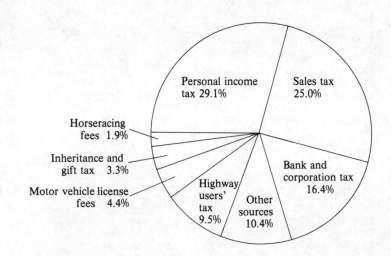

Note that: $360°(.291) = 104.76°$ $360°(.019) = 6.84°$

     $360°(.095) = 34.2°$ $360°(.164) = 59.04°$

     $360°(.044) = 15.84°$ $360°(.250) = 90.0°$

     $360°(.033) = 11.88°$ $360°(.104) = 37.44°$

b. Percentage of persons naturalized under various legal provisions.

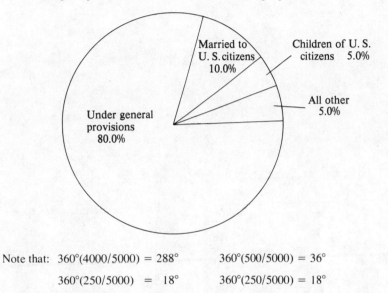

Note that: $360°(4000/5000) = 288°$ $360°(500/5000) = 36°$

     $360°(250/5000) = 18°$ $360°(250/5000) = 18°$

c. Revenue spending.

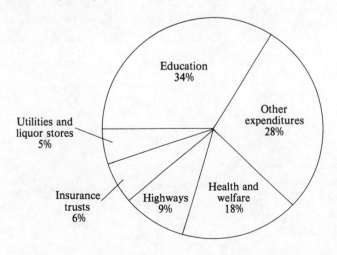

Note that:

$$360°(.34) = 122.4° \qquad 360°(.09) = 32.4°$$
$$360°(.18) = 64.8° \qquad 360°(.05) = 18.0°$$
$$360°(.06) = 21.6° \qquad 360°(.28) = 100.8°$$

d. Average pupil-teacher ratio by country.

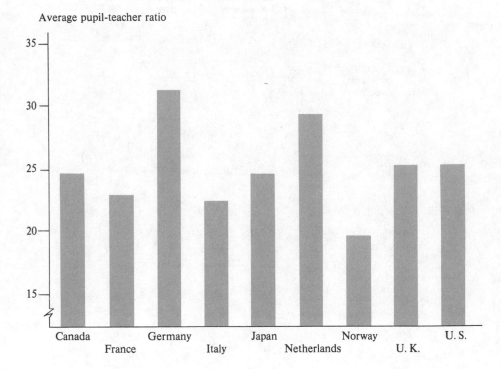

Average pupil-teacher ratio

3. a. 4.5–5.5   **b.** 4.95–5.05

    c. 12–13   **d.** 11.5–12.5

    e. 75–85

4.

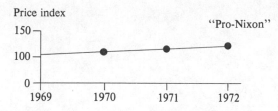

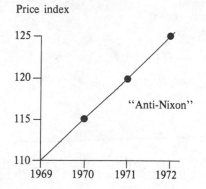

5.

Apparent interval (pounds of pasta)	Real interval (pounds of pasta)	Midpoints	f	Cum. f	Cum. rel. f
91–95	90.5–95.5	93	1	40	1.00
86–90	85.5–90.5	88	4	39	.98
81–85	80.5–85.5	83	7	35	.88
76–80	75.5–80.5	78	8	28	.70
71–75	70.5–75.5	73	6	20	.50
66–70	65.5–70.5	68	6	14	.35
61–65	60.5–65.5	63	4	8	.20
56–60	55.5–60.5	58	3	4	.10
51–55	50.5–55.5	53	1	1	.02

6. Frequency polygon:

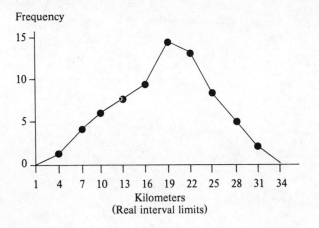

Histogram:

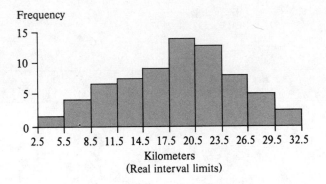

Cumulative relative frequency graph:

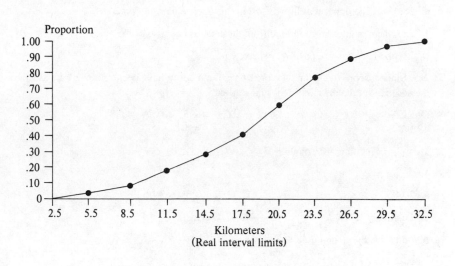

Proportion

7. **a.** positive skew

   b. symmetrical

   c. symmetrical, bimodal (one would expect sex differences here)

   d. negative skew

   e. positive skew (in fact, this distribution might look like a reverse J-curve)

   f. symmetrical

   g. positive skew

8. Although the subclassifications are exhaustive (reading material is either fiction or nonfiction), they are not mutually exclusive (for example, a biography also is nonfiction).

Chapter 3 (pages 45–49)

1. "Lower than average" (that is, than the arithmetic mean). Notice that both the median and modal allowances are equal to Jane's; it is the mean that best serves her purpose.

2. a. Positive skew

    b. 10–14

    c. The modal interval is *less than 10*.

    d. 19.5–24.5

    e. 28.5–? (The upper real limit is lost in this open-ended interval.)

    f. The mean cannot be computed from these data, because of the open-ended intervals.

    g. Mean (The data are positively skewed.)

3. No. Since a score *below* and a score *above* the median have been removed, the location of the score 50 in the distribution is undisturbed.

4. Old $\Sigma X = 5000$, old $N = 100$; new $\Sigma X = 4895$, new $N = 98$; new $\overline{X} = 4895/98 = 48.95$.

5. Mean $= 6$; median $= 5$; mode $= 8$.

6. 4.0

7. a. 20,000   **b.** 19,500–39,500

    c. Positively skewed   **d.** Mean

    e. $53,674.76   Computed as follows:

$$39,500 + 20,000\left(\frac{(512/2) - 110}{206}\right) = 53,674.76$$

    f. $60,554.69   We show you, below, how we arrived at the answer. Your entries for the last three columns may differ from ours (because of a different arbitrary origin) but you should come to the same conclusion.

House price	X'	f	d	d'	fd'
160,000–179,000	169,500	4	120,000	6	24
140,000–159,000	149,500	8	100,000	5	40
120,000–139,000	129,500	18	80,000	4	72
100,000–119,000	109,500	25	60,000	3	75
80,000–99,000	89,500	41	40,000	2	82
60,000–79,000	69,500	100	20,000	1	100
40,000–59,000	49,500	206	0	0	0
20,000–39,000	29,500	110	–20,000	–1	–110
		$N = 512$			$\Sigma fd' = 283$

$$\overline{X} = \frac{20,000(283)}{512} + 49,500 = 60,554.69$$

8. a. Mode = 57 (the crude mode)

b. 60.8   Arrived at as follows:

Speed	X'	f	fX'
80–84	82	3	246
75–79	77	9	693
70–74	72	17	1224
65–69	67	21	1407
60–64	62	30	1860
55–59	57	37	2109
50–54	52	17	884
45–49	47	8	376
40–44	42	5	210
35–39	37	2	74
30–34	32	1	32

$$N = 150$$
$$\Sigma fX' = 9115$$

$$\overline{X} = 9115/150$$
$$= 60.8$$

9. Median = 106.8   Computed as follows:

$$105.5 + 5\left(\frac{170/2 - 79}{24}\right) = 106.8$$

Mean = 107.5   Computed as follows:

IQ's	X'	f	d	fd
146–150	148	2	45	90
141–145	143	1	40	40
136–140	138	4	35	140
131–135	133	7	30	210
126–130	128	10	25	250

(Table continues on next page.)

IQ's	X'	f	d	fd
121–125	123	11	20	220
116–120	118	14	15	210
111–115	113	18	10	180
106–110	108	24	5	120
101–105	103	25	0	0
96–100	98	16	− 5	− 80
91–95	93	16	−10	−160
86–90	88	8	−15	−120
81–85	83	6	−20	−120
76–80	78	5	−25	−125
71–75	73	3	−30	− 90
		$N = 170$		$\Sigma fd = 765$

$$\overline{X} = 765/170 + 103$$
$$= 107.5$$

Again, notice that your entries may differ from these, although your answer should be the same as ours.

10. The median would not be affected by such a correction. However, because every score enters into the computation of the mean, this measure of central tendency would take a different value. The mode probably would not be affected, although this depends on the number of scores involved.

11. **a.** Negative skew

   b. Symmetrical

   c. Positive skew

   d. Symmetrical and bimodal

12. *More people* does not necessarily mean *crowd*. Rinkydink Airlines may be flying only a few more passengers a day than its nearest competitor.

13. Did we catch you napping here? The answer depends on whether or not the annual salary increases were of the same size. If annual increases were equal, the mean and median would be equal. If the annual increases were larger over time, the mean would be more profitable. If, conversely, annual increases were smaller each year, the median gives the advantage.

14. Although we are sure that you would not engage in so unethical a strategy, you can "efficiently" improve the group median by focusing on the group of persons slightly below the median, forsaking all others. The median would not be affected by improve-

ment of persons who already score above it. Improvement of the lowest scorers would change the median only if those improvements were *very* large. Because the mean takes the size of every score into account, the "efficient" (but unconscionable) strategy would be to work with the subset of the total group that you believe will make the greatest gains—again, largely ignoring the others.

15. To calculate the mean, median, and mode, the following table is constructed:

Weights	Interval real limits	X'	f	Cum. f	d	d'	fd'
90–94	89.5–94.5	92	10	80	20	4	40
85–89	84.5–89.5	87	17	70	15	3	51
80–84	79.5–84.5	82	11	53	10	2	22
75–79	74.5–79.5	77	6	42	5	1	6
70–74	69.5–74.5	72	5	36	0	0	0
65–69	64.5–69.5	67	3	31	− 5	−1	− 3
60–64	59.5–64.5	62	7	28	− 10	−2	− 14
55–59	54.5–59.5	57	13	21	− 15	−3	−39
50–54	49.5–54.5	52	8	8	− 20	−4	−32
			$N = 80$				$\Sigma fd' = 31$

$$\bar{X} \text{ (using formula 3.3)} = \frac{5(31)}{80} + 72 = 73.9$$

$$\text{median} = 74.5 + 5 \, \frac{80/2 - 36}{6} = 77.8$$

crude mode $= 87$ (Interval midpoint of interval containing largest number of persons)

Chapter 4 (pages 65–66)

1. Regardless of the measure of dispersion employed, variability in typing scores increases slightly from midsemester to semester end:

	Midsemester	Semester end
Absolute range	65	80
Interquartile range	15.18	19.37
Standard deviation	12.32	15.82

On the next page is the table we constructed to calculate the standard deviations.

		Midsemester					Semester end				
Words per minute	X'	f	d'	fd'	d'²	f(d'²)	f	d'	fd'	d'²	f(d'²)
105–109	107						1	7	7	49	49
100–104	102						1	6	6	36	36
95–99	97						2	5	10	25	50
90–94	92						2	4	8	16	32
85–89	87	1	6	6	36	36	3	3	9	9	27
80–84	82	1	5	5	25	25	4	2	8	4	16
75–79	77	2	4	8	16	32	6	1	6	1	6
70–74	72	3	3	9	9	27	7	0	0	0	0
65–69	67	4	2	8	4	16	7	–1	–7	1	7
60–64	62	7	1	7	1	7	6	–2	–12	4	24
55–59	57	9	0	0	0	0	4	–3	–12	9	36
50–54	52	9	–1	–9	1	9	3	–4	–12	16	48
45–49	47	6	–2	–12	4	24	1	–5	–5	25	25
40–44	42	4	–3	–12	9	36	1	–6	–6	36	36
35–39	37	2	–4	–8	16	32	1	–7	–7	49	49
30–34	32	1	–5	–5	25	25	1	–8	–8	64	64
25–29	27	1	–6	–6	36	36					
		$N = 50$		$\Sigma fd' = -9$		$\Sigma (fd'^2) = 305$	$N = 50$		$\Sigma fd' = -15$		$\Sigma (fd'^2) = 505$

Midsemester standard deviation:

$$s = 5 \sqrt{\frac{50(305) - 81}{2500}} = 5 \sqrt{6.068} = 12.32$$

Semester-end standard deviation:

$$s = 5 \sqrt{\frac{50(505) - 225}{2500}} = 5 \sqrt{10.01} = 15.82$$

2. Since Q_1 and Q_3 represent the 25th and 75th percentiles, respectively, a change in an extreme score would not affect their values. However, since *every* score in a distribution enters into the computation of the standard deviation, a change in an extreme score—or any score—affects this measure of dispersion.

3. a. .1587   **b.** .0228   **c.** .8664

4. You would expect greater similarity in the standard deviations. The interquartile range depends solely on the location of two scores in a distribution and, consequently, is rather unstable from one sample of a population to another (for example, from one random half of Boy Scouts to the other). The standard deviation, rather than relying only on two scores, considers every score in the distribution; this measure of dispersion has greater stability. Thus, you would expect that the standard deviations of the two random halves of Boy Scouts would be more similar than the two interquartile ranges.

5. Relative to its average price, Klimon cheese is more variable:

$$CV \text{ (Klimon)} = 20 \qquad CV \text{ (Etsaton)} = 10$$

6. In this case, the standard deviation will be of more interest to the instructor. Although the two sections are comparable on *mean* mechanical aptitude, there is four times the variability in section A than in section B; there will be more high-aptitude as well as low-aptitude students in section A. Because of this greater heterogeneity in Section A, the instructor may consider supplemental materials, further grouping, special instruction, and so on.

Just for fun, assume that these data for each section are normally distributed about their respective means. For each section, supply the corresponding raw score for each z score we have provided below, and notice the difference in dispersion:

Section A
( )	( )	( )	( )	( )	( )	( )
-3	-2	-1	0	+1	+2	+3

Section B
( )	( )	( )	( )	( )	( )	( )
-3	-2	-1	0	+1	+2	+3

7. Clearly, this distribution is negatively skewed. With $\bar{X} = 95$ and 100 the maximum score possible, there is not sufficient area beyond the mean to accommodate a standard deviation of 12. Thus, the distribution necessarily must tail off below the mean. (We leave it to you to guess how difficult this spelling test was.)

8. Before comparing the variability of these two groups, we must employ Scheppard's correction for the grouped data:

$$s \text{ (corrected)} = \sqrt{(6.1)^2 - \frac{12^2}{12}} = 5.0$$

After this correction, we see that school A has greater variability in knowledge of current affairs.

Chapter 5 (pages 75–77)

1. a.

Horse	Rank	Horse	Rank
Valiant	1	Sally	6
Crystal	3	Danny	7.5
Uma-san	3	Cavallo	7.5
Dandy Billy	3	Paisano	9
Copper Bars	5	Wrangler	10

b. Definitely. For example, Valiant and Copper Bars are four ranks apart, with Valiant having two blue ribbons more than Copper Bars. Now compare Copper Bars to Sally. Copper Bars has five blue ribbons more than Sally, although these two horses are only one rank apart. Thus, although Copper Bars and Sally are considerably more similar in rank than Valiant and Copper Bars, the actual raw-score difference between Copper Bars and Sally is more than twice the raw-score difference between Valiant and Copper Bars.

2. a. Unimodal (mode = 10)   **b.** 9

    c. 8   **d.** 70

    e. 30.5   **f.** 5%

3. a. Ted's z scores are: verbal, + .5; quantitative, + .75; logical + 1.00   He performed relatively highest on logical reasoning, and relatively lowest on verbal. Ted's total score is + 2.25.

    b. $z = + 1.50$; PR = .93

    c. $z = 3.0$ This score is unusual, indeed. A z score of this magnitude is located so far above the mean of a normal distribution that almost all of the scores fall below it—.9987 of them, to be exact. (Appendix II brings us to this conclusion. We add the value in column 2 to .50, which gives us .9987—or the area of the distribution below a z of 3.00.)

    d. $z = - 3.78$

Chapter 6 (pages 88–90)

1. The problem is called a *restricted* or *truncated* range. When all pupils in an elementary school are included in the calculation of the correlation coefficient, there is a great deal of variation in both age and reading ability. In this case, you expect to find a positive correlation between these two variables—six-year-old children (first-graders) typically do not read as well as ten-year-old children (fifth-graders) who, in turn, typically do not read as well as 13-year-old children (eighth-graders). But when you focus on one grade (that is, when you restrict the range of grade), this vast variation is reduced appreciably and, consequently, the magnitude of the correlation is reduced. (This, incidentally, is why one often hears of the low correlation between performance in graduate school and entrance exam scores. Applicants are selected, in part, because they score high on these entrance exams—a policy that reduces variation in this variable and, hence, its correlation with subsequent performance in graduate school.)

Although variation is reduced with restriction of range, there is a chance for something else to happen, as well: the sign of the correlation coefficient may change. You would not be completely surprised if you were told that, in a particular grade, the older children were found to do less well than the other children. This might result, for example, from the fact that these older children were held back from promotion to the next grade level. Thus, the correlation might be slightly negative. But when taken in the context of the entire elementary school, the correlation is positive. The overly-simplistic figure below illustrates why. Here, we have reading ability along the vertical axis, and age along the horizontal axis—a continuous variable ranging from, say, 6 to 13. Each of the eight oblong circles represents a separate scatterplot of scores conveniently—and slightly unrealistically—corresponding to each grade in this elementary school. The scatterplot in the lower-left corner represents the first-grade data, while the scatterplot in the upper-right corner represents the eighth-grade data. As you can see, each scatterplot reflects a slightly negative relationship between age and reading ability (that is, when range is restricted on age). However, when taken as a whole, these data reflect a positive relationship between the two variables. Older children in a particular grade may be doing less well *relative to their grade,* yet still performing better than the children in earlier grades.

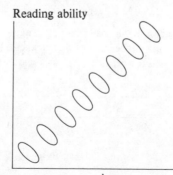

Reading ability

Age

2. $\rho = 1 - \dfrac{6(34.5)}{10(10^2 - 1)} = 1 - \dfrac{207}{990} = .79$

3.

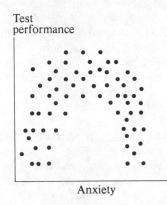

Anxiety

These data do not meet the assumption of linearity that underlies the Pearson r.

4. a. positive   **b.** negative
   c. negative   **d.** positive

5. a. Both likely are a function of increase in population: as a population grows in size, more newspaper subscriptions are sold, as well as more washing machines (or any other popular product).

   b. Although a continuously debated issue, this correlation likely arises from systematic differences in experience had by members of different socioeconomic statuses. Two (of many) sources of these differences are quality of schooling and patterns of socialization in the home and in the community.

   c. Three factors are: (1) the person who has relatively lengthy absences from work due to sickness may tend to quit his job because of this sickness; (2) this person may tend to be fired, or released, from his job because of his sickness; and (3) this person may die.

6.

X	Y	X^2	Y^2	XY
1	5.1	1	26.01	5.1
2	11.4	4	129.96	22.8
3	8.6	9	73.96	25.8
4	2.6	16	6.76	10.4

(Table continues on next page.)

X	Y	X^2	Y^2	XY
5	9.0	25	81	45
6	6.7	36	44.89	40.2
7	6.0	49	36	42
8	19.4	64	376.36	155.2
9	12.8	81	163.84	115.2
10	20.7	100	428.49	207
11	70.9	121	5026.81	779.9
12	63.1	144	3981.61	757.2

$\Sigma X = 78$ $\Sigma X^2 = 650$ $\Sigma XY = 2205.8$
$\Sigma Y = 236.3$ $\Sigma Y^2 = 10375.69$
$\bar{X} = 6.5$ $\bar{Y} = 19.69$

$$r = \frac{\dfrac{2205.8}{12} - (6.5)(19.69)}{\sqrt{\left(\dfrac{650}{12} - 6.5^2\right)\left(\dfrac{10375.69}{12} - 19.69^2\right)}}$$

$$= \frac{55.83}{\sqrt{(11.92)(476.94)}} = \frac{55.83}{75.40} = .74$$

7. a.

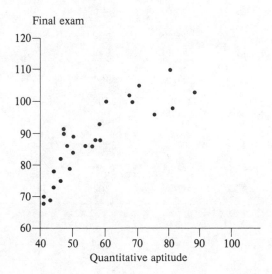

Final exam
Quantitative aptitude

b.

X	Y	X^2	Y^2	XY
68	41	4624	1681	2788
84	50	7056	2500	4200
70	40	4900	1600	2800
110	80	12100	6400	8800
80	49	6400	2401	3920
98	81	9604	6561	7938
75	46	5625	2116	3450
103	88	10609	7744	9064
100	68	10000	4624	6800
86	54	7396	2916	4644
88	58	7744	3364	5104
102	67	10404	4489	6834
69	43	4761	1849	2967
73	44	5329	1936	3212
86	56	7396	3136	4816
100	60	10000	3600	6000
82	46	6724	2116	3772
90	47	8100	2209	4230
96	75	9216	5625	7200
84	50	7056	2500	4200
88	57	7744	3249	5016
93	58	8649	3364	5394
91	47	8281	2209	4277
105	70	11025	4900	7350
89	50	7921	2500	4450
78	44	6084	1936	3432
86	48	7396	2304	4128

$\Sigma X = 2374$ $\Sigma Y = 1517$ $\Sigma X^2 = 212144$ $\Sigma XY = 136786$

$\bar{X} = 87.93$ $\bar{Y} = 56.19$ $\Sigma Y^2 = 89829$

$$r = \frac{\dfrac{136786}{27} - (87.93)(56.19)}{\sqrt{\left(\dfrac{212144}{27} - 87.93^2\right)\left(\dfrac{89829}{27} - 56.19^2\right)}}$$

$$= \frac{125.36}{\sqrt{(125.50)(169.68)}} = \frac{125.36}{145.93} = .86$$

Appendix I (pages 102–103)

1. –2

2. $3X$

3. X^2

4. 8

5. 24

6. –5

7. 6

8. $-X$

9. 1

10. $X^2/4$

11. 4/6 or 2/3

12. 14.4

13. 4.5

14. –4

15. 3.10805

16. 9.48683

17. .307409

18. 35.0714

19. 4

20. 6, 7.8, .8, 45, 0, 20, 100

21. a. $3\sqrt{X}$

　b. $\dfrac{X + Y}{\frac{1}{2}}$　or　$2(X + Y)$

　c. $\dfrac{X}{5}$ [2(.05)]

INDEX